Where the Universe Came From

How Einstein's Relativity Unlocks the Past, Present and Future of the Cosmos

NEW SCIENTIST

New Scientist

First published in Great Britain by John Murray Learning in 2017
An imprint of John Murray Press
A division of Hodder & Stoughton Ltd,
An Hachette UK company

This paperback edition published in 2021

1

A CIP catalogue record for this title is available from the British Library

B format ISBN 978 1 52938 183 2
eBook ISBN 978 1473 62960 8

Typeset by KnowledgeWorks Global Ltd.

Printed and bound in Great Britain by Clays Ltd, Elcograf S.p.A.

John Murray Press policy is to use papers that are natural, renewable and recyclable products
and made from wood grown in sustainable forests. The logging and manufacturing processes
are expected to conform to the environmental regulations of the country of origin.

John Murray Press
Carmelite House
50 Victoria Embankment
London EC4Y 0DZ

Nicholas Brealey Publishing
Hachette Book Group
Market Place, Center 53, State Street
Boston, MA 02109, USA

instantexpert.johnmurraylearning.com

Also available
as an ebook

Where the Universe Came From

How Einstein's Relativity Unlocks the Past,

Present and Future of the Cosmos

Contents

Series introduction

New Scientist's *Instant Expert* books shine light on the subjects that we all wish we knew more about: topics that challenge, engage enquiring minds and open up a deeper understanding of the world around us. *Instant Expert* books are definitive and accessible entry points for curious readers who want to know how things work and why. Look out for the other titles in the series:

Contributors

Editor-in-chief: Alison George is *Instant Expert* editor for *New Scientist*.

Editor: Stephen Battersby is a physics writer and consultant for *New Scientist*.

Articles in this book are based on talks at the 2016 *New Scientist* masterclass 'Relativity and beyond' and articles previously published in *New Scientist*.

Academic contributors

Michael Duff is Emeritus Professor of Theoretical Physics at Imperial College London, and a pioneer of supergravity. He wrote "A theory of 'everystring'" in Chapter 8.

Pedro Ferreira is Professor of Astrophysics at the University of Oxford. He works on the general theory of relativity and on the nature of dark matter and dark energy. He wrote "The Light Bends" in Chapter 1.

John Gribbin is an astrophysicist and science writer. He is a visiting fellow in astronomy at the University of Sussex, UK, where he investigates the age of the universe. He wrote "A very special theory" and "A theory of some gravity" in Chapter 2.

Martin Hendry is Professor of Gravitational Astrophysics and Cosmology at the University of Glasgow, UK, and a specialist in gravitational wave astronomy. He wrote about what space-time curves into, in Chapter 2, and "Are there particles of gravity in these waves?" in Chapter 4.

Dan Hooper is an associate scientist at the Fermi National Accelerator in Batavia, Illinois and Assistant Professor of Astronomy and Astrophysics at the University of Chicago. He wrote "Shedding light on dark matter" and "What is dark matter" in Chapter 6.

Sabine Hossenfelder is a research fellow at the Frankfurt Institute for Advanced Studies, where she researches quantum gravity. She co-wrote "Strangely familiar" in Chapter 6.

Eugene Lim is a theoretical cosmologist at King's College London. His interests range from string theory to the role of quantum information in the cosmos. He wrote "The odd couple" in Chapter 8.

Andrew Pontzen is a lecturer at University College London, where he researches galaxy formation and computational cosmology. He wrote "A relatively brief history" in Chapter 1.

Marika Taylor is Professor of Theoretical Physics at the University of Southampton, UK, and an expert on black holes. She wrote "Studying the invisible" in Chapter 3.

Milena Warzeck is a historian of science who focuses on the social and political context of modern science. She wrote "The Relativity Deniers" in Chapter 1.

Thanks also to the following writers:

Anil Ananthaswamy, Jacob Aron, Michael Brooks, Marcus Chown, Stuart Clark, Daniel Cossins, Amanda Gefter, Lisa Grossman, Naomi Lubick, Govert Schilling, Joshua Sokol, Colin Stuart, Richard Webb, Jon White.

Introduction

Look behind that unmistakable flare of fluffy white hair and you'll find a whole new view of the world, where time and space are joined, mass becomes energy and the fabric of the universe is revealed ... and then torn to shreds.

Albert Einstein forged his theories of relativity more than 100 years ago in what was one of the greatest achievements of the human mind, and yet today we are still finding out what they really mean. One ramification of relativity is our growing understanding of the life of the cosmos, from its beginning in the Big Bang through many stages of expansion. Another is the uncanny thing called dark energy, which dominates the universe, and was heralded in Einstein's own calculations in the 1920s.

In 2016 we saw perhaps the most relativistic moment in history, when scientists detected gravitational waves coming from the collision of two black holes after being stretched out by the expansion of space-time over a billion years. Soon, gravitational wave detectors and radio telescopes could begin to explore the nature of the event horizon, the point of no return at the edge of a black hole, to find out whether relativity holds up even at these extremes. Meanwhile, a frenzy of theoretical ideas is emerging from the clash between relativity and quantum mechanics, with superstrings, quantum triangles and other strange notions competing to provide a deeper description of reality. Even Einstein must be surpassed some day.

This *New Scientist Instant Expert* guide gathers together the thoughts of leading physicists and the best of *New Scientist* magazine to bring you up to date with Einstein's concept of relativity and the impact it has had on our conception of the universe.

Stephen Battersby, Editor

I

The roots of relativity

In 1915 a patent clerk in Switzerland came up with an idea that transformed our conception of space and time. The clerk was Albert Einstein (1879–1955) and the idea was general relativity. This chapter describes the path that led to his momentous discovery.

A relatively brief history

First, we should get this out of the way: Einstein was not a lone genius. His contribution was immense, but it did not exist in a vacuum.

The story starts when Scottish physicist James Clerk Maxwell (1831–79) achieved one of the great unifications of physics. In the 1860s he took many separate theories of how magnetic and electric fields work, and showed that they could all be described in one set of equations. Then he made a remarkable prediction: the combined fields responsible for electric and magnetic forces could form a type of wave that would travel at the speed of light. By the end of the nineteenth century it was accepted that this was no coincidence: light itself was composed of these 'electromagnetic waves'.

Strangely, the equations said that the waves always travel at an identical speed, regardless of how their source is moving or how fast you as an observer are moving. That didn't seem quite right. If I throw anything forwards from a moving vehicle, it flies faster than if I throw it from a standing start. Why should light be any different?

Based on this logic, people started to look for some kind of variation in the speed of light. The most famous attempt was an experiment in 1887 by the American physicists Albert Michelson (1852–1931) and Edward Morley (1838–1923), who tried to detect light changing its speed as the Earth spins and swings around the Sun. They took a beam of light, split it in two and sent it along two arms at a 90-degree angle – expecting to find slight differences in the time it took to go along each arm, depending on how the set-up was oriented relative to the motion of the Earth. But, however hard they looked, they always found that light moves at the same speed.

In 1895 the Dutch mathematician Hendrik Lorentz (1853–1928) came up with a way to understand the constant speed of light. He developed a set of rules that relate what you see when moving to what you would see when standing still (see Chapter 2). The rules involve what he regarded as a fictional time: if you are moving along at high speed, then you need to use this fictional time, different from the time that would be measured by a normal clock. Using this mathematical trick, everything seems to make sense and the speed of light appears the same for everyone.

Time warp

Five years later, the French scientist Henri Poincaré (1854–1912) wrote an essay ('La mesure du temps') in which he wondered why we thought of time so rigidly. Tellingly, Lorentz thought of warping time as just a mathematical trick – but Poincaré (without referring explicitly to Lorentz) pointed out that in future it might be necessary to abandon the idea of a unique notion of physical time. This was a philosophical leap that helped free Einstein to formulate relativity.

On the subject of philosophy, a second nudge that would influence Einstein's later work came from the Austrian physicist and philosopher Ernst Mach (1838–1916). In his 1883 textbook *The Science of Mechanics*, Mach asserted that we should never talk about how something is moving in absolute terms – only about how it is moving relative to something else.

Finally, the groundwork was laid for Einstein. In his 1905 paper 'On the electrodynamics of moving bodies', he started with two assumptions:

1 The laws of physics should be the same when we write them relative to any frame of reference moving at constant velocity.

2 We need to take Maxwell's equations seriously – any ray of light moves in any such frame of reference with exactly the same velocity.

About Albert Einstein

Albert Einstein was born in Ulm, south-west Germany, on 14 March 1879, the second child of Hermann Einstein, founder of an electrical engineering company, and his wife Pauline. The family, who were non-observant Ashkenazi Jews, soon moved to Munich where Albert later went to school.

Aged 17, Einstein entered the Swiss Federal Polytechnic School in Zurich, to train as a teacher of physics and mathematics. It was here that he met his fellow student Mileva Maric, whom he married in 1903. In 1987 newly discovered letters between the two revealed that they had a child out of wedlock in 1902, though the fate of this girl is unknown – she may have been adopted or died in infancy. The couple later had two sons, Hans and Eduard, before separating and eventually divorcing in 1919, when he married his cousin Elsa Löwenthal (née Einstein).

After graduating, Einstein spent a frustrating two years trying to secure a teaching position and eventually ended up working at the Swiss Patent Office. It was here, and in his spare time, that he made his early discoveries, including the remarkable series of papers of his *annus mirabilis* in 1905 (see 'The miraculous year' in Chapter 2). This led to his appointment in 1908 as a lecturer at the University of Bern, Switzerland, which was swiftly followed by a professorship at the University of Zurich. By 1914 he was a professor at the University of Berlin, where he remained for almost two decades until the political situation in Germany changed and the Nazi government began to prohibit Jews from holding teaching positions at universities. In 1933 he gave up his citizenship and emigrated to America, finding refuge at the Institute of Advanced Study in Princeton, New Jersey, where he remained until his retirement.

Einstein is known not only for his remarkable discoveries. He was also an ardent music lover, a pacifist, a champion of civil rights and a supporter of Zionism. He died of an aneurism in 1955, aged 76, and his ashes were scattered in an unknown location – although his brain was preserved (see later in this chapter).

Relatively special

In a few short pages, Einstein was able to derive a cornucopia of results that we now know as special relativity. Many of these results had appeared before, but now they were unified and given a clear physical interpretation. It was clear, for example, that time dilation is real, not fictional: moving clocks really should slow down. Perhaps because Lorentz and Poincaré had laid so much groundwork, Einstein's 1905 special relativity seems not to have raised too much controversy. Certainly, it did not make anything like the popular impression of his later general theory, which would require over a decade's more work to reach.

One of the first developments towards that goal seemed inauspicious: the Polish-German mathematician Hermann Minkowski (1864–1909) found a neat way of explaining special relativity. He coined the idea of space-time – that space and time are intertwined. You can think about a map of how things unfold in time and space: at the bottom of the map you have the far past, at the top the far future, while left or right mark different places. Minkowski realized that, when you are moving, you point in a different direction in space-time: instead of straight up the page, you tilt to the left or right. Mathematically, it is very much like a rotation that interchanges some of your space for time and some of your time for space. This abstract view correctly generates the results of special relativity in a beautifully streamlined way.

But Einstein recognized that special relativity was limited. It correctly relates different frames of reference only if they are moving at constant velocities. He was also worried about gravity. The best theory of gravity at the time was Newton's. Newton, like Maxwell, was a unifier: he showed that the same force that keeps us glued to the Earth's surface also stops the Moon flying off into space and keeps the Earth orbiting around the Sun. The resulting theory works extremely well, but it involves a kind of instantaneous pull: somehow just the presence of the Earth below exerts a force on us. Even on cosmic scales, you feel the tug of all the galaxies around you at any moment. That doesn't sit comfortably with special relativity, where nothing can travel instantaneously; to make it all fit together, nothing should move faster than the speed of light, not even forces.

The principle of equivalence

Einstein took a first step to bringing gravity into his theory in 1907, by formulating what is now called the principle of equivalence. He pointed out that when you are falling, in some sense there is no gravity. If you look around you, other things that are falling seem motionless – because everything is falling at the same rate. This is what happens on the International Space Station: it's not that the astronauts have escaped Earth's gravitational pull, it's that the space station is constantly falling towards Earth while,

FIGURE I.I Albert Einstein in 1904, aged 25

say, astronaut Tim Peake was also constantly falling towards Earth at the same rate. (The station never crashes to Earth because it is also moving horizontally at high speed.)

Inspired by Mach's earlier philosophy, Einstein's genius was to take the bold step of insisting that, for any experiment carried out in the microcosm of a space station, the result should be just the same as if gravity simply did not exist. That is the principle of equivalence.

It is bizarre that Einstein's theory of gravity should be based on deep thought about situations where the very force we are talking about disappears. No surprise, then, that it needed a great deal of mathematical development to turn the idea into a theory that could make meaningful predictions. In 1913

Einstein had started to play with Minkowski's idea of space-time as a tool. He found that he could get the right result for how objects move in a gravitational field by assuming that space-time is warped, and that objects try to follow the shortest route through this curved space-time, but he wasn't able to show what makes it curve in the first place.

By now, Einstein was struggling with the maths. For a frenetic few months in 1915 he was corresponding with many different people, in particular the German mathematician David Hilbert (1862–1943). Einstein's and Hilbert's work became so intertwined that it is unclear who exactly wrote down the field equations first. But there is no doubt that Einstein was the driving force. Finally, in November 1915 in his general theory of relativity, he was able to describe how space-time gets curved by the presence of mass, energy and pressure:

$$G_{\mu\nu} = \frac{8\pi G}{c^4} T_{\mu\nu}$$

There is great richness embedded in these few characters. Within six months of finding the field equations, Einstein was writing papers on gravitational waves, a hundred years before they were finally directly detected (see Chapter 4). Black holes were also predicted shortly after the theory was published (see Chapter 3).

Other consequences took much longer to emerge. In 1949 the Austrian-American mathematician and philosopher Kurt Gödel (1906–78) mounted an attack on relativity. A lover of absurdity, Gödel was able to show that general relativity permitted time travel into the past. This is anathema to physicists: if it is possible to travel back into our own past, what is to stop us changing it? Any science fiction fan will tell you that this is not advisable.

Wormholes and more

Gödel's example required the whole universe to be spinning, which it is not (as far as we can tell). But in 1988, the physicists Mike Morris and Kip Thorne found another route to time travel. They showed that wormholes – short cuts from one part of space-time to another – could, in principle, be opened if an exotic type of energy could be produced by an advanced civilization. Once open, they can be used to zip across space and time. While the prospects are remote, time travel is seemingly permitted by Einstein's equations; this still provokes plenty of heated discussion between physicists.

In the meantime, there is plenty to be working on. It has only recently become possible to solve Einstein's equations on computers, which has opened up a whole new way of exploring the bizarre behaviour of black holes and other exotic objects. Combined with the detection of gravitational waves, we finally seem to be getting to grips with the theory and its implications – a process that has taken a hundred years. But we must remember that the richness of relativity tells us not only about the genius of Einstein but of his predecessors, his contemporaries and the many people who have worked to figure out what it all means.

'The most lucid, not to mention entertaining, proponent of Einstein's ideas has always been Einstein himself.'

Stephen Hawking, *A Stubbornly Persistent Illusion* (2008)

Einstein in his own words

In 2010 Albert Einstein's original handwritten manuscript 'The Foundation of the Generalized Theory of Relativity' was put on display for the first time in its entirety at Israel's Academy of Sciences and Humanities in Jerusalem.

Einstein wrote the 46-page paper in 1916 – three years before the theory's first major confirmation of general relativity during an eclipse. The paper mentions the potential test of the theory, as well as its prediction for the perihelion of Mercury's orbit, which had, until general relativity, remained an anomaly. He also commented in the paper that it remained 'an open question whether the theory of the electromagnetic field in conjunction with that of the gravitational field furnishes a sufficient basis for the theory of matter or not'.

Writing in 1916, Einstein didn't yet know of the two other forces that would have to be taken into account – the weak and strong nuclear forces – but his question was profound and remains an open one today. Legions of physicists are trying to answer a similar question, as they seek to unite general relativity with quantum mechanics in an ultimate theory of everything.

There's a particular thrill that comes from reading Einstein in his own words (digitized versions of this paper and others can be found online). His unique philosophical style is at times deceptively simple, full of useful thought experiments and always questioning even our most basic assumptions about reality. In 1921 he was awarded the Nobel Prize in physics for his 'services to theoretical physics, and especially for his discovery of the law of the photoelectric effect'.

The light bends

How did Einstein's theory hold up to real-world tests?

The theory of relativity is often considered a triumph of pure intellect, one of the most elegant of fundamental physical theories. But elegance and intellect mean nothing in physics if they don't match our observations of nature.

For more than 200 years, Newton's theory of gravity had passed this test with flying colours. At its core was the law of universal gravitation: the force of gravity between any two objects is proportional to each of their masses and inversely proportional to the square of their distance apart. Newton's law was used to predict the motion of planets in our solar system with remarkable accuracy. Such was its power that in 1846 the French astronomer Urbain Le Verrier (1811–77) used it to predict the existence of Neptune.

There was only one case where Newton's theory did not give the right answer. Le Verrier found that Mercury's orbit

FIGURE 1.2 Albert Einstein in 1947, aged 68

drifted by a tiny amount – less than one-hundredth of a degree over a century – relative to what would be expected from Newton's theory. This puzzling discrepancy remained until 1916, when Einstein showed that his general theory of relativity would lead to the observed drift in Mercury's orbit. General relativity passed its first test almost immediately.

Einstein also predicted that a massive object such as the Sun should distort the path of light: in effect, the curved geometry of space should act as a lens and focus the light (see Figure 1.3). (In fact, Newton's theory also predicts that light will curve, but only half as much as in general relativity.)

Lucky eclipse

On 11 August 1999, the skies above Albert Einstein's birthplace in the city of Ulm in Germany darkened as the Moon eclipsed the Sun. It was a fitting tribute to the man who transformed our picture of the natural world – and an unlikely one, too. Although a total eclipse occurs roughly every 18 months somewhere in the world, at a given location the gap between successive eclipses is about 350 years. What were the chances, then, that the greatest scientist of the twentieth century should be commemorated by the last total eclipse of the millennium? But perhaps we should not be too surprised about the coincidence; for Einstein, eclipses have always been lucky.

Take the pivotal role played by a similar total eclipse nearly a hundred years ago in establishing Einstein's general theory of relativity. Einstein's paper, smuggled out of Germany during the First World War, reached the British physicist Arthur Eddington (1882–1944) at Cambridge.

It was Eddington who realized that the total eclipse of 29 May 1919 that would occur above the island of Príncipe off the coast of West Africa would offer a golden opportunity to test one of general relativity's central predictions.

Eddington's team travelled to Príncipe and duly photographed their quarry. They wanted to observe a bright cluster of stars called the Hyades as the Sun passed in front of them and, to blot out the solar glare, Eddington needed a total eclipse. If Einstein's theory was correct, the positions of the stars in the Hyades would appear to shift by about 1/2000th of a degree.

Eddington first took a picture of the Hyades at night from Oxford. Then, on 29 May 1919, he photographed the Hyades from Príncipe as they lay almost directly behind the Sun during the eclipse. It took a long time to detect this bending of light – the shift in the position of the stars was so small – but in September 1919 Eddington finally announced that Einstein was right. Comparing the two measurements, Eddington found that the shift was as Einstein had predicted. The result turned Einstein into an international superstar.

Perhaps this was another piece of luck for Einstein. The rigour of Eddington's result is somewhat controversial today; some have suggested that the light-bending effect was actually too small for Eddington to have discerned clearly, and had he not been so receptive to Einstein's theory, he might not have reached such strong conclusions so soon.

Since then, Einstein's theory has passed many more tests. One prediction is that as light climbs out of the warped space-time near a massive object, its wavelength should be stretched, or redshifted. In 1959 the American physicists Robert Pound (1919–2010) and Glen Rebka (1931–2015) measured this gravitational redshift in their laboratory at Harvard. We have ample evidence of black holes, large and small (see Chapter 3). And, of course, in 2016 physicists at the LIGO collaboration finally detected gravitational waves (see Chapter 4) – the travelling distortions of space-time that Einstein had predicted a century earlier.

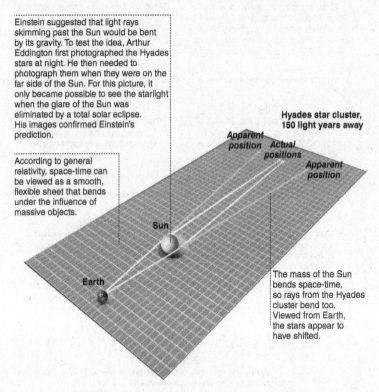

Einstein suggested that light rays skimming past the Sun would be bent by its gravity. To test the idea, Arthur Eddington first photographed the Hyades stars at night. He then needed to photograph them when they were on the far side of the Sun. For this picture, it only became possible to see the starlight when the glare of the Sun was eliminated by a total solar eclipse. His images confirmed Einstein's prediction.

According to general relativity, space-time can be viewed as a smooth, flexible sheet that bends under the influence of massive objects.

Hyades star cluster, 150 light years away

Apparent position   Actual positions

Apparent position

Sun

Earth

The mass of the Sun bends space-time, so rays from the Hyades cluster bend too. Viewed from Earth, the stars appear to have shifted.

FIGURE 1.3 Light is bent by curved space-time

Space-timeline

1905
Einstein presents his **special theory of relativity** in a paper titled 'On the electrodynamics of moving bodies'.

1915
Einstein presents his **field equations of general relativity** to the Prussian Academy of Sciences in Berlin.

1929
Edwin Hubble and others show that far-off galaxies are moving away from us – the first hint of an expanding **'big bang'** universe. Einstein drops his cosmological constant.

1922
Alexander Friedmann finds a solution to Einstein's equations that describes a **uniformly expanding universe**. Five years later, Georges Lemaître independently comes up with the same finding.

1948
Theorists predict that, if the universe is expanding from a hot and dense beginning in a 'big bang', it should have left behind an afterglow: the **cosmic microwave background**.

1964
The cosmic microwave background is discovered as noise in a radio antenna, kicking off relativity's 'golden age'.

1998
Studies of far-off supernovae reveal that the **universe's expansion is accelerating**. Einstein's cosmological constant is revived as one possible cause.

1989
NASA launches COBE, a satellite to study the cosmic microwave background. It reveals a **largely homogeneous radiation field**, supporting the idea of an inflationary big bang.

2000s
More detailed studies of the cosmic microwave background support the picture of a cosmos that began in an inflationary Big Bang dominated by **dark matter** and **dark energy**.

1916

Einstein uses general relativity to predict the existence of **gravitational waves**, ripples in space-time produced when anything with mass accelerates.

1917

Einstein introduces an extra term into his equations, the **cosmological constant**, to balance out gravity and produce a static universe that is neither expanding nor contracting.

1921

Einstein wins the Nobel Prize 'for his services to theoretical physics, and especially for his discovery of the law of the photoelectric effect.'

1919

Arthur Eddington observes the Sun's mass bending light during an eclipse over the island of Príncipe – the **gravitational lensing effect** predicted by Einstein.

1972

X-ray emissions from a body known as X-1 in the constellation Cygnus provide the first evidence for a star's collapse into a stellar-mass **black hole**.

1974

Russell Hulse and Joseph Taylor discover a pair of neutron stars whose orbits are slowing exactly as if they were losing energy by emitting **gravitational waves**.

1980

Alan Guth and others propose that the Big Bang universe was smoothed out by undergoing a period of breakneck expansion in its first instants – **inflation**.

1974

Stephen Hawking shows theoretically that quantum effects can cause black holes to evaporate, emitting **Hawking radiation** – posing the question of what happens to the information they swallow.

2016

Advanced LIGO detects gravitational waves from colliding black holes.

Was there anything special about Einstein's brain?

When Einstein died, the pathologist, keen to find out the source of Einstein's intellectual power, removed the brain from the body and then dissected and photographed it. The brain was initially something of a disappointment, being slightly smaller than average. However, in recent decades, the images of Einstein's brain have given rise to new insights. A 1999 study showed that Einstein's parietal lobe, the area of the brain associated with mathematics and spatial reasoning, was 15 per cent wider than in a normal brain. The National Museum of Health and Medicine in Chicago has even made an Einstein Brain Atlas app, using more than 350 digitized versions of these slides, so that researchers can delve into the great man's grey matter. According to a 2012 study published in the neurology journal *Brain*, Einstein's remarkable intelligence might be attributable to his prefrontal cortex, which is responsible for speech as well as imagining events and simulating their consequences. Compared with the average brain, his is dramatically expanded. The researchers also noticed a large knob on his motor cortex, representing Einstein's early practice playing the violin.

Some sections of the brain are unaccounted for, raising the possibility that there's a piece of it in your grandparents' attic.

2

About space and time

The theory of relativity encompasses two separate theories devised by Einstein in the early twentieth century: special and general. Here is a primer on both.

A very special theory

Einstein's 1905 special theory of relativity transformed our understanding of space and time.

Einstein painted a strange new picture of the universe, where observers see moving clocks run slow, moving rulers shrink, and moving masses become even more massive. All these shocking consequences stem from two simple ideas: the constant nature of the speed of light, no matter who measures it; and the principle of relativity, which holds that the same laws of physics apply for all observers moving in straight lines at constant speeds.

To understand why, it is traditional to imagine a train (see Figure 2.1). Bob, who is travelling on the train, sets up a source of light in the middle of his carriage, which sends off two pulses of light in opposite directions. From Bob's point of view, the pulses will reach the end walls of the carriage simultaneously. But, standing on the platform, Alice sees something else. To her, the speed of each of the two light pulses is still exactly the same as the speed measured by Bob; but while the light pulses are travelling, the train moves forward. Alice sees the rearward-moving pulse hit the back wall of the carriage before the forward-moving pulse hits the front wall. So two events that are simultaneous for one person are seen to occur at different times by someone else. Simultaneity is relative.

If observers cannot agree on simultaneity, they cannot agree on the outcome of measurements involving time. This gives us the phenomenon known as time dilation. Bob takes a clock on the train – a perfect clock which measures time by how long it takes light to bounce back and forth between two mirrors, at right angles to the train's motion. The passage of a light pulse from one mirror to the other and back again is one tick of the clock.

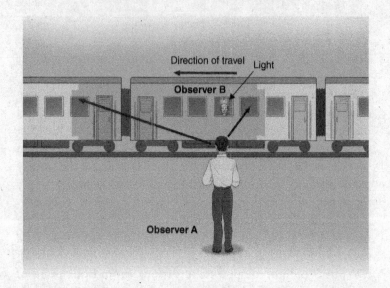

FIGURE 2.1 According to relativity, simultaneous events for one observer may seem to occur at different times for another.

Bob knows the speed of light and the distance between the mirrors, so he knows the time taken for one tick.

Constant speed

But for Alice on the platform, the clock and its mirrors move forward, so the path travelled by the light pulse is along two sides of a triangle. This path is longer than the straight distance between the two mirrors at rest. Because the speed of light is a constant, one tick of the moving clock takes longer, according to Alice, than one tick of an identical clock sitting on the platform.

It is important to realize that the situation is symmetrical. By the principle of relativity, Bob can regard the train as at rest and the platform as moving, and make the same calculation to infer that Alice's clock is running slow. There is no paradox here once we remember that simultaneity is relative. We can't compare the readings of the separated clocks 'at the same instant' until we decide what 'at the same instant' means, and our two observers have different views on this.

Since the light pulse in the moving clock travels along the hypotenuses of two right-angled triangles, it is easy to calculate the size of the time dilation. If v is the speed of the moving clock, and c is the speed of light, time stretches by a factor of $1/\sqrt{\left(1-v^2/c^2\right)}$. This quantity, known as the Lorentz factor, appears in many relativistic calculations.

Time dilation

While time dilation can be seen most clearly in this special kind of clock, it is a real effect that applies to all moving clocks and processes. Experiments with fast-moving but short-lived subatomic particles show that their lifetimes are indeed extended by the Lorentz factor.

So much for time; how about space? Suppose there is a long stick lying on a table in the train. Alice can measure the length of the stick by counting how many ticks of the platform clock it takes for the stick to pass a certain point on the platform. But according to Bob, her clock is running slow so, compared with his measurement, Alice's measurement of length is shrunk, by the same factor of $1/\sqrt{\left(1-v^2/c^2\right)}$.

This FitzGerald-Lorentz contraction also applies to the train, and to Bob. Everything is shrunk in the direction of its motion. Of course, for motion much slower than light, the factor is small: even for a supersonic jet moving at Mach 2 – twice the speed of sound at sea level – the contraction is only two parts

23

The miraculous year

The year 1905 was Einstein's *annus mirabilis* – the year in which he, a dapper 26-year-old, produced four papers that changed the world. On 9 June he published his work on the photoelectric effect, making a major leap forward in the field of quantum physics when he showed how energy comes in discrete packets. It was this work, not relativity, that won him a Nobel Prize. A month later came another paper – a theory about Brownian motion, the random movement of particles in liquids and gases. On 26 September his work on special relativity was published, and then on 21 November he introduced what is possibly the most famous equation: $E = mc^2$.

in 1 million million. As an object moves faster relative to an observer, it is increasingly foreshortened and its clocks tick ever more slowly. At the speed of light, an object's length in the direction of motion is zero, and its time stands still.

Because length and time depend on your frame of reference, velocities don't add up in the way you might expect. Say Bob's train is going at velocity $v1$, and he fires a bullet ahead at velocity $v2$, as measured from the train. Alice on the platform will see the bullet flying by not at $v1 + v2$, but more slowly. The velocity she sees is $(v1+v2)/(1+(v1v2)/c^2)$.

This means that the bullet – or any other object – can never be seen to move faster than c by any inertial observer (anyone moving at constant velocity). For example, if the train moves at $0.75c$ and the bullet is fired at $0.75c$, Alice sees the bullet moving at $0.96c$, not at $1.5c$.

But what has happened to the bullet's energy? Energy has to be conserved for both Bob and Alice. The gun gives the bullet extra energy, but from Alice's point of view its velocity does not increase enough to account for that energy input. Kinetic energy is $1/2 \, mv^2$ (where m is mass) so, if the velocity has not gone up enough, the mass must have increased.

In other words, moving objects have more mass than they do when at rest. Einstein calculated that the mass of a moving object is equal to its mass when stationary, multiplied by the familiar Lorentz factor.

That famous equation

The increase in mass works out to be equal to the increase in energy divided by c^2, and Einstein inferred that the mass at rest is equivalent to an energy E/c^2 – in other words, $E = mc^2$.

This equivalence between energy and mass is true for all forms of energy. It provided a fundamental answer to the puzzle of radioactivity. The French physicist Pierre Curie (1859–1906) had found in 1903 that 1 gram of radium emits more than 400 joules of heat per hour. Where was it all coming from? Einstein argued that when a radioactive element disintegrates, some of its mass is turned into energy in line with the equation $E = mc^2$. If it were fully converted, a gram of radium would contain enough energy to keep a one-kilowatt electric heater burning for 2,850 years.

All the strange predictions of special relativity have been borne out by experiment. It was by using special relativity that the British physicist Paul Dirac (1902–84), in 1928, developed a description of how electrons behave. His relativistic version of quantum mechanics provides the understanding of the behaviour of electrons in atoms, and the way in which they occupy stable 'shells' around the nucleus – the basis of chemistry. So every chemical process in your body is witness to the world where time, space, energy and mass are all relative.

A mathematical marriage

Both space and time lose their status as absolute properties of nature in Einstein's theory. But the German mathematician Hermann Minkowski showed how to combine
space and time into something more fundamental.

Think of how an object like a broom handle can appear
to have a different length, depending on its orientation
(see Figure 2.2). Side on, you see its full length. Head on,
it appears to have no length at all. At intermediate angles
it is foreshortened. Minkowski pointed out that all the
strange results of special relativity could be understood

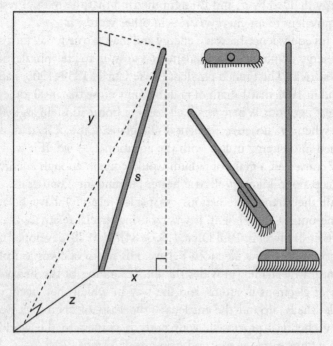

FIGURE 2.2 Minkowski's illustration of the theory of relativity

as views of an object with different orientations in four-dimensional space-time.

Giving an object such as a broom handle a four-dimensional 'length' means taking into account the instants of time at which we observe the ends of the handle. If we make the observations at different times, the handle has an extent in time as well as space.

Because light travels at 300,000 kilometres per second, one second of time is equivalent to 300,000 kilometres of space. Of course, space and time are clearly different things in our experience, and that is reflected in the mathematics. In ordinary 3D space, the length of the broom handle (s) can be given in terms of its extent in three dimensions x, y and z, by the formula

$$s^2 = x^2 + y^2 + z^2$$

But in Minkowski's 4D space-time, the time part isn't added to the formula – it is subtracted. The true four-dimensional extent of an object is

$$s^2 = x^2 + y^2 + z^2 - c^2 t^2$$

This space-time interval, s, is always measured to be the same by all inertial observers. Although different observers will describe the length of a particular object in different ways, and will measure the speed of a clock attached to it in different ways, the correct combination of space and time properties belonging to the object gives an unambiguous measure of its extent in space-time.

Einstein was initially reluctant to take this idea seriously, but he later realized that Minkowski's idea provided the key to a general theory of relativity, incorporating gravity (see 'A theory of some gravity' below).

A theory of some gravity

Albert Einstein's general theory of relativity is one of the towering achievements of twentieth-century physics and it made him a household name. Published in 1916, his theory explains that what we perceive as the force of gravity in fact arises from the curvature of space and time.

The insight that set Einstein on the path to the general theory of relativity came when he realized that a person trapped in a freely falling elevator does not feel the force of gravity. They will float, weightless, able to push themselves from floor to ceiling with ease. We have now seen astronauts in this situation – pencils hanging in mid-air, liquids that refuse to pour and so on – but Einstein had to imagine it. His genius was to grasp its significance. If the downward acceleration of the falling lift can precisely cancel out the force of gravity, then that force and acceleration are equivalent to each other. This is the principle of equivalence (see Figure 2.3).

To gain an idea of how powerful this idea is, imagine the elevator replaced by a closed laboratory. It is a rocket-powered lab, being accelerated through space by a constant force. Everything in the lab falls to the floor. Physicists can carry out experiments there to measure the downward force on objects, but they will not be able to tell whether that is due to acceleration or gravity.

One clever physicist might decide to send a beam of light across the room, at right angles to the acceleration. In the time it takes the light to cross the room, the target wall has speeded up and moved sideways relative to the beam. Looking at the spot of light on the wall, it will seem as though the beam of light is bent. So can the physicist distinguish acceleration from gravity in this way? No, says Einstein: by the principle of equivalence, a light beam must also be bent by gravity.

FIGURE 2.3 Einstein's equivalence principle: gravity and acceleration produce identical forces and no experiment can distinguish between them

Four dimensions

Einstein puzzled over this idea before coming up with a mathematical theory that explained light bending and much more besides (see Chapter 1). His picture of the universe cast aside the everyday notion of empty space and replaced it with an almost tangible continuum in four dimensions – three of space and one of time. This is based on Minkowski's idea of space-time as a way to understand special relativity – but while that space-time was flat, the continuum of general relativity can be curved. Both energy and pressure can bend space-time, but in practice the main cause of curvature is the mass-energy of matter.

FIGURE 2.4 The fabric of reality: massive objects bend space-time

Four dimensions are hard to visualize, so instead imagine a two-dimensional rubber sheet stretched tightly across a frame. Roll a marble across the sheet: it makes a tiny dent and rolls in a straight line. Add a bowling ball, which bends the sheet downwards. Now the marble follows a curved path. That is Einstein's model for the force of gravity: objects follow the most direct path, called a geodesic, through curved space-time. This is true for a marble, a planet or a beam of light.

The most visually striking consequence of this is the gravitational lens, where a cluster of galaxies or some other concentration of mass bends and focuses light from a distant object to make two or more images on the sky – occasionally smearing it out into a shining circle known as an Einstein ring (see Figure 2.5).

Where gravity is weak, relativity gives essentially the same answers as Newton's theory: it acts like a force between objects that falls off as the square of the distance. But in a strong field, new effects become noticeable. At the distance of Mercury from the Sun, this means a shift in the planet's orbit, which was a puzzling discrepancy until explained by general relativity (see 'The light bends' in Chapter 1).

The most extreme deviation from Newton's laws is the black hole. A black hole bends space-time around itself so much that it becomes closed off from the rest of the universe. By the

FIGURE 2.5 An Einstein ring – these rings are the artefact of a lens made of warped space-time

rubber-sheet analogy, it creates a deep funnel from which not even light can escape. In the very centre of a black hole, at a point known as a singularity, density becomes infinite and both the analogy and the equations break down.

A model of the cosmos

The theory also deals naturally with the whole universe. But at first, when Einstein used it to make a mathematical model of the cosmos, he ran into a problem. In 1917 the received wisdom was that the universe was static, whereas the equations of general relativity insisted that it must be either expanding or contracting. The only way Einstein could hold a model universe still was to add an extra term to the equations, called the cosmological constant. A dozen years later, observers, led by the pioneer Edwin Hubble (1889–1953) in California, found that the universe is expanding. In the relativistic picture, the rubber sheet is being continually stretched in all directions. This not only means that distant galaxies are receding, but that the wavelength of light stretches out as it travels, causing redshift.

The same analogy helps us to visualize how gravitational waves originate. When a lump of matter vibrates, it sends out ripples through the sheet, and these ripples set other lumps of matter vibrating. Gravitational waves are very weak, but finally in 2016 scientists detected them directly (see Chapter 4). Einstein's general theory of relativity is now established beyond doubt as the best one we have to explain gravity and the universe at large.

How can I understand the concept of relativity?

Space and time used to be so simple. You trundled around reasonably freely in the three dimensions of the one, and experienced occasional heartache at the remorseless forward march of the other. *C'est la vie*.

Or is it? Einstein revolutionized our perceptions a century ago when, in his theories of relativity, he first forbade anything in the cosmos from travelling faster than the speed of light, and then bundled both space and time into one unified space–time that can be warped by gravity. The contortions introduced by Einstein's special and general theories make intervals in both space and time dependent on from where we measure them. Two observers with flashlights in fast-moving trains might both measure the other to have flashed their flashlight first – and both would be right from their own point of view.

The blockbuster film *Interstellar* (2014) is based on premises that Einstein made technically plausible, if not yet technologically feasible: that by travelling close to the speed of light, or moving in an intense gravitational field such as that of a black hole, we age more slowly than the people we leave behind on Earth (see Figure 2.6). We do not need to travel that far to see less dramatic effects

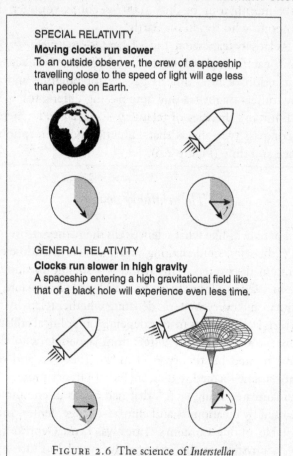

SPECIAL RELATIVITY

Moving clocks run slower
To an outside observer, the crew of a spaceship travelling close to the speed of light will age less than people on Earth.

GENERAL RELATIVITY

Clocks run slower in high gravity
A spaceship entering a high gravitational field like that of a black hole will experience even less time.

FIGURE 2.6 The science of *Interstellar*

of relativity in action. Astronauts on the International Space Station age a little less because of the velocity at which they travel, and a little more for enjoying less of the gravity of Earth. The effects do not quite cancel out. Velocity wins, leaving each ISS astronaut who completes a six-month tour of duty 0.007 seconds younger than someone who stayed on Earth.

Relativity can seem full of paradoxes unless we first think carefully about how our own motion affects our perception of how time is passing for others – and also how others might see our time passing differently.

Einstein's theories of relativity – special and general – encompass two effects that influence our perception of space and time (Figure 2.6).

The relativity deniers

When people dislike what science tells them, they resort to conspiracy theories, mud-slinging and plausible pseudoscience – as Einstein discovered. 'This world is a strange madhouse,' he wrote in 1920 in a letter to his friend Marcel Grossmann. 'Every coachman and every waiter is debating whether relativity theory is correct. Belief in this matter depends on political affiliation.'

Einstein received many letters from laypeople who claimed to have refuted his theory. And in the 1920s an anti-relativity movement formed, which included physics professors and Nobel laureates, using tactics that had much in common with those used by creationists and climate-change deniers today.

Notable among Einstein's critics was Ernst Gehrcke (1878–1960), a physicist at the Imperial Technical Institute in Berlin. Like many experimentalists, he felt uncomfortable with a theory that changed the fundamental concepts of space and time. In 1921 he argued that giving up the idea of absolute time threatened to confuse the basis of cause and effect. He first raised his objections in scholarly journals. But after a key prediction of general relativity was confirmed during the eclipse of 1919 (see Chapter 1), Einstein was transformed into a media star and the debate acquired a much broader public impact.

The controversy in Germany intensified in August 1920 with a series of public lectures against Einstein at the Berlin Philharmonic Hall. They included a talk by Gehrcke, who repeated the arguments he had been raising unsuccessfully for years, as well as an impassioned speech by the German scientist and anti-Semitic activist Paul Weyland (1888–1972), who had organized the series. This event prompted Einstein to consider leaving Germany.

Gehrcke was in touch with a range of Einstein's opponents, from astronomers to philosophers to schoolteachers, including physics Nobel laureates Johannes Stark and Philipp Lenard. An organization called the Academy of Nations was formed, whose title and letterheaded paper contrived to give it the aura of a scholarly academy. In fact, it served as an international network of Einstein's opponents. Its founder was the Swedish-American scientist Arvid Reuterdahl (1876–1933), then dean of engineering and architecture at the University of St Thomas in St Paul, Minnesota.

Concerned that science was becoming too specialized, the Academy of Nations aimed to reconnect different branches of knowledge by integrating scientific findings into a unified, religious account of nature. To Reuterdahl, nothing better symbolized the modern specialization and incomprehensibility of science than relativity, and almost half the Academy of Nations' founding declaration was polemics against Einstein's theory. The smear campaign orchestrated by this seemingly reputable scholarly academy shows that Einstein was up against implacable enemies.

'Crazy vagary'

The American section of the Academy included some prominent figures, such as astronomer Thomas See (1866–1982) of the US Naval Observatory at Mare Island, California. In the early 1920s he published several articles in which he accused Einstein of plagiarism and denounced the theory as a 'crazy vagary'.

Reuterdahl was eager to establish contact with Einstein's opponents all over the world, and he approached Gehrcke in 1921 with the idea of setting up a German branch of the Academy. Gehrcke's first recruits were German physicists who argued that there was no need for relativity because classical physics could explain all astronomical observations. Philosophers, engineers and physicians joined too, and even a retired major general.

Why did this ramshackle alliance between laymen and scientists emerge? One of the more noble reasons was that Einstein's opponents were genuinely concerned about the future of science. The increasingly mathematical nature of theoretical physics collided with a view that science should be simple mechanics, comprehensible to an educated layperson. Relativity was a threat – a theory that 'only 12 wise men' could comprehend, as Einstein was reported in the *New York Times* in 1919 to have said (Figure 2.7). The increasing role of advanced mathematics seemed to disconnect physics from reality.

The 1920s were also an unsettling decade for Germany, with hyperinflation, political upheavals, and radical cultural developments such as Dadaism and Expressionism. In an uncertain world, some felt that science should be relied upon to provide firm ground.

Some opponents attacked Einstein the person – the democrat, the pacifist, the Jew. Others turned to anti-Semitic conspiracy theories. Reuterdahl wrote in 1923: 'Our trouble in America is that all scientific journals are closed to the anti-relativists through

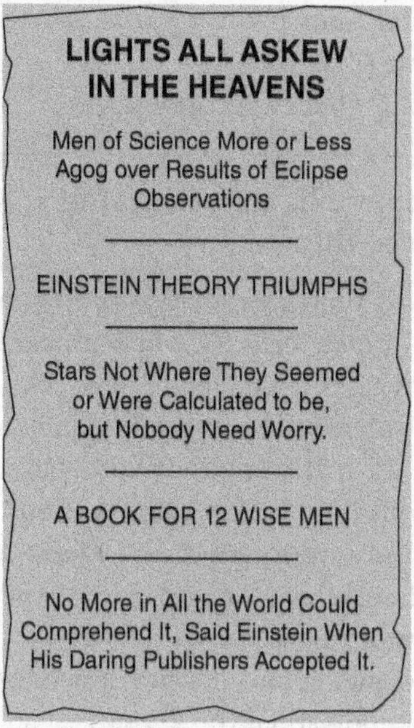
LIGHTS ALL ASKEW
IN THE HEAVENS

Men of Science More or Less
Agog over Results of Eclipse
Observations

EINSTEIN THEORY TRIUMPHS

Stars Not Where They Seemed
or Were Calculated to be,
but Nobody Need Worry.

A BOOK FOR 12 WISE MEN

No More in All the World Could
Comprehend It, Said Einstein When
His Daring Publishers Accepted It.

FIGURE 2.7 Headlines in the *New York Times* reporting the confirmation of Einstein's general theory of relativity that followed the eclipse expedition to Príncipe, 1919 (see Chapter 1)

Jewish influence. The daily press is almost entirely under the control of the Jews.'

By the mid-1920s the relativity deniers faced overwhelming resistance, and most refrained from taking a public stance against relativity. Many simply gave up, and the Academy of Nations ceased to serve as the central organization campaigning against Einstein. Yet some opposition to relativity lingers even today. The website Conservapedia claims that relativity is 'heavily promoted by liberals' and lists 32 reasons why the theory is wrong. But at least anti-relativism today has less academic participation and a much lower profile than it had in the 1920s. Probably only a small minority of waiters today are debating whether relativity theory is correct.

When people talk about curving space-time, what does it curve into?

This relates to a question that cosmologists have wrestled with over the past century, often phrased as 'If space is expanding, what is it expanding into?' The short answer is that it doesn't have to be expanding into anything.

We often use a 2D metaphor for the expansion of space, thinking about how, for example, dots on the surface of a balloon appear to move apart as the balloon inflates. Here, the 2D surface of the balloon is the counterpart of our three dimensions of space. Of course, we can see that the balloon is expanding into another dimension. But we could still tell something about the 2D surface and how it curves if we were embedded within it. By looking at the properties of those dots on the surface and how curved lines and angles behave as the balloon expands, we can distinguish its surface from that of a flat piece of paper, without needing to worry about any higher dimensions.

This is what we call the intrinsic curvature of the balloon's surface. In a similar way, the changes in space-time caused by heavy objects or gravitational waves can be described in terms of the intrinsic curvature of space-time. The only dimensions we need are the three of space and one of time.

However, although higher dimensions are unnecessary, they may exist. In some speculative theories of physics, the universe is a curved membrane ('brane' for short) that floats in a higher-dimensional space.

3
Black holes

In the last century black holes have moved from being a disputed idea to playing a central role in our understanding of the cosmos. What are they, and what happens when one of them swallows matter?

Warped space-time

It was while he was serving in the German army on the Russian front, in the winter of 1915–16, that the physicist Karl Schwarzschild (1873–1916) sent Albert Einstein some papers. He had solved Einstein's equations of general relativity for the first time, and shown what happens to space-time inside and outside a massive object – in this case, a perfectly spherical, non-spinning star. Einstein was thrilled.

He wouldn't have been so thrilled with a prediction that eventually emerged from Schwarzschild's work. Make a star massive or dense enough and it could develop a gravitational pull so great, and warp space-time so much, that even light would not escape.

Just months after his exchange with Einstein, Schwarzschild was dead. It was left to others to work through the details of these curious objects, known as Schwarzschild singularities. Chief among them was a young Indian physicist named Subrahmanyan Chandrasekhar (1910–95). In 1930 he boarded a steamer from India to the UK, where he was to take up a scholarship at the University of Cambridge. Whiling away the 18-day voyage, he worked on the properties of white dwarf stars. He found that, if they had more than 1.4 times the Sun's mass, they would implode under their own gravity, forming a Schwarzschild singularity.

This did not go down well. At a meeting of the Royal Astronomical Society in 1935, the eminent astrophysicist Arthur Eddington declared that 'there should be a law of nature to prevent a star from behaving in this absurd way'. In 1939, Einstein himself published a paper to explain why Schwarzschild singularities could not exist outside the minds of theorists.

Collapsing stars

The impasse remained until the 1960s, when physicists such as Roger Penrose (1931–) proved that black holes – a term coined at about this time and adopted by the astrophysicist John Wheeler (1911–2008) – were an inevitable consequence of the collapse of massive stars. At the centre of a black hole, physical quantities such as the curvature of space-time would become infinite, and the equations of general relativity would break down. Not only that, but a black hole's interior would be permanently hidden behind its event horizon, the surface of no return for light. That in turn meant that nothing happening in the interior could influence events outside, because no matter or energy could escape.

Although we cannot see a black hole directly, in 1970 astronomers observing a compact object in the constellation Cygnus saw jets of X-rays consistent with theoretical predictions of radiation streaming from hot matter spiralling towards an event horizon (see Figure 3.1). Since then, our appreciation of black holes' reality has only grown.

The characteristics of black holes remain hotly disputed, however – not least for what they say about general relativity's

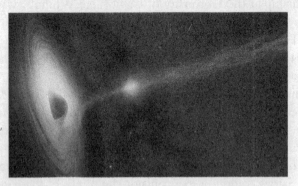

FIGURE 3.1 An artist's impression of a black hole surrounded by a swirling disc of material falling on to it

failure to mesh with quantum theory, which explains behaviour at the atomic and subatomic level. Tensions rose in the 1970s, when physicists Jacob Bekenstein (1947–2015) and Stephen Hawking (1942–) showed that black holes must have a temperature. Bodies with a temperature have an associated entropy and, in quantum mechanics, entropy – a measure of disorder – implies the existence of a microstructure. Einstein's equations, meanwhile, describe black holes as smooth, featureless distortions of space-time. Hawking also showed that quantum effects in and around the event horizon imply that the black hole should steadily evaporate, emitting a stream of what we now call Hawking radiation.

The firewall paradox

If a black hole eventually dwindles to nothing, what happens to the stuff that falls in? At a fundamental level, matter and energy carry information, and quantum mechanics says that information cannot be destroyed. Perhaps the information encoded slips out with the Hawking radiation, but this idea runs into another problem: it leads to the black hole being surrounded by a firewall of blazing, energetic particles, again something general relativity forbids.

This firewall paradox is still a hot topic. One emerging and tantalizing suggestion is that the smooth fabric of Einsteinian space-time results from particles inside and outside the event horizon being linked quantum mechanically, via structures known as wormholes. In August 2015 Hawking set out another idea, suggesting that information is never actually swallowed by a black hole. Instead, it persists at its event horizon in a form that is garbled and hard to decode. A month later, Nobel laureate Gerard't Hooft (1946–) of Utrecht University in the Netherlands suggested that, when matter and energy fall in, their information just bounces back.

Studying the invisible

Measuring the properties of black holes is not straight-forward. By definition, they cannot emit light, so you cannot just point a telescope at one and observe it. But you can see its gravitational effects.

In 1971 an object called Cygnus X-1 was postulated to be a stellar-mass black hole (created by the gravitational collapse of a massive star) because of the strong gravitational effect it had on its companion star. Three years later came the discovery of Sagittarius A*, a supermassive black hole in the centre of our own galaxy. The orbits of the stars around it show that something extremely heavy is there, four million times the mass of the Sun.

There are other ways to detect black holes. Although the event horizon cannot emit light, many black holes are surrounded by accretion discs of gas spiralling into the black hole. These are hot and emit radiation across a range of frequencies – from radio waves through visible light to X-rays. And if a black hole is spinning, it can spew out huge jets of matter.

Soon, we might get much closer to really seeing a black hole. A project called the Event Horizon Telescope aims to image Sagittarius A* and other supermassive black holes (see the interview on photographing giants, below). And we are going to learn a lot about black holes from gravitational waves (see Chapter 4), perhaps enough to find out what really happens at the event horizon.

Some sidestep such problems by returning to arguments reminiscent of Eddington's and Einstein's denial of black holes. In 2014, Laura Mersini-Houghton of the University of North Carolina, Chapel Hill, argued that massive stars cannot collapse to black holes – the emission of Hawking radiation during the collapse stops the star from ever getting that far. So there are no event horizons and no singularities.

Few subscribe to that view. Instead, the firewall paradox has opened up a new front in the struggle to unite general relativity and quantum mechanics. In that tussle, there is a sense that the successful theory will be closer to quantum theory than general relativity, given the overwhelming success of quantum theory in explaining all the forces of nature besides gravity. That might not have pleased Einstein, who was troubled both by black holes and what he saw as quantum theory's excesses. Black holes could end up being the prediction that ate his theory.

When black holes collide

Meet the challenger: a black hole new to the neighbourhood, with a mass 140 million times that of the Sun. This plucky upstart is 35 times more massive than the black hole that holds court at the centre of our Milky Way. And now, make way for the current champion: a black hole with a mass of 18 billion suns.

For front-row seats to this cosmic boxing match, you'll want (cautiously) to approach OJ 287, the core region of a galaxy 3.5 billion light years away. Here, the smaller black hole orbits its larger rival. With every trip around, it falls closer, on track to be swallowed up in about 10,000 years. But in the meantime, it is putting up an admirable fight.

Even though the system is so far away, OJ 287 releases enough energy to appear about as bright in the sky as Pluto. We have been capturing it on photographic plates since the 1880s, but it first caught the eye of Mauri Valtonen at Finland's Tuorla Observatory in Turku almost a century later. His team noticed that, unlike other galactic centres, which brighten and dim sporadically, this one seemed to keep to a tight schedule. About every 12 years, it has an outburst. The gap between outbursts grows shorter by about 20 days each cycle. In the decades since we noticed the pattern, we have gone a long way towards working out why.

OJ 287 is a window into what must have happened in galaxies all over the universe. Galaxies grow by eating their own kind, and almost all of them come with a supermassive black hole at the centre. When two galaxies unite, their black holes will either tussle until one is ejected with a gravitational kick, or they will spiral together and eventually merge into an even bigger black hole.

In OJ 287, the smaller black hole is en route to merging with the larger one. The larger one is also growing from a surrounding disc of gas and dust that slowly swirls down the drain. Each time the smaller black hole completes an orbit, it comes crashing through this disc at supersonic speeds. That violent impact blows bubbles of hot gas that expand, thin out and then unleash a flood of ultraviolet radiation. You could stand 36 light years away and tan faster than you would from the Sun on Earth.

While these outbursts may be impressive, the black holes' orbital dance emits tens of thousands of times more energy than the undulations in space-time called gravitational waves (see Chapter 4). The smaller black hole has no chance of escape. Those waves leach energy from the binary orbit, bringing the pair closer together and making each cycle a little shorter than the last.

In 2015 the Laser Interferometer Gravitational-wave Observatory (LIGO) in the US offered a preview of the endgame of OJ 287 in miniature. Twice it heard gravitational waves from the final orbits of black-hole pairs, in which each black hole was a few dozen times the size of the Sun, and then the reverberations of the single one left behind.

Because its black holes are so massive, the ultimate collision at the heart of OJ 287 will be too low frequency for LIGO to hear. But the outcome will be much the same. Where once two black holes from two separate galaxies tussled, one black hole will remain, smug and secure at the centre.

Interview: Photographing giants

Dan Marrone is an astronomer at the University of Arizona Steward Observatory. He is part of the Event Horizon Telescope programme, which aims to take the first picture of a black hole.

A black hole, by definition, is black. So how are you going to take a picture of one?

If you look right at the black hole, it should look quite dark, as very little light escapes. But just around the edge of it you see a bright ring, which is due to the photons that just missed going into the black hole and skimmed around the edge of it a couple of times. This light is what we think we will be able to detect with the Event Horizon Telescope (EHT).

The EHT is a 'whole Earth' telescope. How does it work?

In radio astronomy, to get a higher resolution than you can from a single telescope, you record signals from many telescopes around the world and add them together. It is as if you have a single telescope almost the size of the Earth.

Which black holes are you targeting?

We're targeting Sagittarius A*, which is the supermassive black hole at the centre of our galaxy, and the black hole at the centre of M87, the biggest galaxy in the Virgo cluster of galaxies. With a telescope the size of the Earth and at the frequencies we are observing, we can just make out black holes of this size.

Every image of a black hole so far has been an artist's impression. Will the real thing match expectations?

The question of creating an image from what we measure is a tricky one. We will most likely represent it as a false colour image, using colours to represent how bright the light is. This image will not be as pretty as an artist's impression. The galaxy blurs the light between us and the black hole, so there are a lot of sharp features we can't possibly see. But any image we get shouldn't disappoint – we are looking at something no one has ever seen before.

What about capturing a moving image – 'Black hole, the movie' as it were?

We can, if there is something orbiting the black hole, as we expect there will be. If there is gas orbiting before it falls into the black hole, this takes between four and 27 minutes, depending on the spin of the black hole. If we look for several days and see changes in the structure, we can represent that as a movie as well.

What are you hoping to learn from this image?

Just being able to take a picture of a black hole, and show this shadow that we expect to be there because the light is not escaping, will be important. Beyond that, we have a lot to learn about the structure of our galaxy's black hole, and what happens to a black hole when it is being starved of material, as Sagittarius A* appears to be.

We also expect to be able to test general relativity, which tells us that the ring of light around the edge of the back hole needs to be perfectly circular. If general relativity is breaking in this very strong field regime, where gravity is at the limits of its power, then this ring of light will not be perfectly circular.

Are black holes hiding other universes?

In the early days of the cosmos, a quirk of space-time may have created wormholes that connected us to a vast multiverse. If borne out, the theory may help explain how supermassive black holes at the centres of galaxies grew so big so fast ... and it would mean that each of these giant holes holds an entire universe within it.

The idea that ours is just one of a staggering number of universes follows from one leading theory of cosmology called eternal inflation. This idea was devised in the 1980s to explain some puzzling observations that the standard Big Bang theory alone could not explain. It holds that far, far beyond our known universe, space-time is expanding exponentially, doubling in volume every fraction of a second. Every now and then, a 'bubble' of space-time drops out of this frenetic expansion to settle into a more sedate rate of growth – as our universe did almost 14 billion years ago. Even after rapid expansion ends here, new baby universes can keep being born elsewhere, spawning a sprawling multiverse.

What would it be like to fall into a black hole?

Lacking any credible eyewitness reports, we have to rely on the theory of relativity to answer this question.

Relativity tells us that from the point of view of any outside observer, you never do quite fall in. As your image gets closer and closer to the event horizon, the hole's gravity begins to tamper with time. Instead of plummeting faster inwards, your image slows down, crawling more and more slowly but never quite getting to the event horizon. At the same time, your image fades to red, then infrared, and finally radio waves of ever-increasing wavelength.

Your own experience would be different, but we don't know exactly how. As you get closer to the hole, the scatter of stars and galaxies across the sky becomes warped and blueshifted to painful brightness. According to your own clocks and perceptions, you reach the event horizon rapidly. You might pass through it unscathed, or you might meet a firewall and be blasted to subatomic particles. If you do survive, then you will probably reach a point close to the centre of the hole where the gradient of gravitational force is so strong it pulls you to pieces, a process known as 'spaghettification'. But according to some theories, you could instead find yourself thrown through a wormhole and into a whole new universe – where the chances are that no one would believe your story.

Alex Vilenkin (1949–), at Tufts University in Medford, Massachusetts, and his colleagues wondered whether they could spot signs of the multiverse. They made a mathematical analysis of the fate of the bubbles formed during inflation. They found that some bubbles would form in a halfway state: still inflating, but at a different rate. Some of them would form within our patch of space, and when that stops inflating entirely these halfway bubbles would be stranded within it. They would appear to us as black holes. Bubbles that formed relatively late would be smaller, and should collapse into standard black holes, with nothing inside apart from an infinitely dense point called a singularity; but earlier bubbles would be bigger, creating larger black holes. Inside each of these, inflating space-time can spawn its own multiverse.

The analysis suggests that our universe should have a distinctive distribution of black holes. The higher the mass, the more of them there should be – up to a critical mass, after which the number should fall. This could help solve a long-standing mystery. Standard astrophysics has difficulty explaining how supermassive black holes became so big so early in cosmic history. In the new theory, big black holes would have been formed in the first moments of the Big Bang as separate bubble universes. These giants could have become the supermassive black holes we see today at the heart of galaxies, including our own Milky Way.

It may also have implications for the black hole information loss paradox, which physicists have battled over for decades. And, of course, our own universe could look like a black hole to physicists in some other universe.

Interview: Stephen Hawking

Stephen Hawking, one of the world's greatest physicists, is famous for his work on black holes. To celebrate his 70th birthday in 2012, New Scientist *conducted an exclusive interview with him via email (he was diagnosed with motor neurone disease aged 21, and can now communicate only via a sensor activated by twitching his cheek). His responses to the questions are followed by* New Scientist's *elaboration of the concepts he describes.*

What has been the most exciting development in physics during your career?

COBE's discovery of tiny variations in the temperature of the cosmic microwave background and the subsequent confirmation by WMAP that these are in excellent agreement with the predictions of inflation. The Planck satellite may detect the imprint of the gravitational waves predicted by inflation. This would be quantum gravity written across the sky.

[The COBE and WMAP satellites measured the cosmic microwave background (CMB), the afterglow of the Big Bang that pervades all of space. Its temperature is almost completely uniform — a big boost to the theory of inflation, which predicts that the universe underwent a period of breakneck expansion shortly after the Big Bang that would have evened out its wrinkles. If inflation did happen, it should have sent ripples through space-time — gravitational waves — that would cause variations in the CMB too subtle to have been spotted so far. Since this interview, the European Space Agency's Planck satellite has been looking for them, so far without success. Other telescopes are aiming to be even more precise than Planck (see Chapter 5).]

Einstein referred to the cosmological constant as his 'biggest blunder'. What was yours?

I used to think that information was destroyed in black holes. But the AdS/CFT correspondence led me to change my mind. This was my biggest blunder, or at least my biggest blunder in science.

[Black holes consume everything, including information, that strays too close. But in 1975, together with the Israeli physicist Jakob Bekenstein, Hawking showed that black holes slowly emit

radiation, causing them to evaporate and eventually disappear. So what happens to the information they swallow? Hawking argued for decades that it was destroyed – a major challenge to ideas of continuity, and cause and effect. In 1997, however, theorist Juan Maldacena developed a mathematical short cut, the 'Anti-de-Sitter/conformal field theory correspondence', or AdS/CFT. This links events within contorted space-time geometry, such as in a black hole, with simpler physics at that space's boundary.

In 2004 Hawking used this to show how a black hole's information might leak back into our universe through quantum-mechanical perturbations at its boundary, or event horizon. The recantation cost Hawking a bet made with fellow theorist John Preskill a decade earlier.]

What discovery would do most to revolutionize our understanding of the universe?

The discovery of supersymmetric partners for the known fundamental particles, perhaps at the Large Hadron Collider. This would be strong evidence in favour of M-theory.

[The search for supersymmetric particles is a major goal of the LHC at CERN. The standard model of particle physics would be completed by finding the Higgs boson, but it has a number of problems that would be solved if all known elementary particles had a heavier 'superpartner'. Evidence of supersymmetry would support M-theory, the 11-dimensional version of string theory that is the best stab so far at a 'theory of everything', uniting gravity with the other forces of nature.]

If you were a young physicist just starting out today, what would you study?

I would have a new idea that would open up a new field.

4
Gravitational waves

We have known for more than a century that gravitational waves must exist, but these ripples in space and time were not finally detected until 2016, when the Laser Interferometer Gravitational-wave Observatory (LIGO) picked up the tell-tale slight stretching and squeezing of space-time caused by the movement of massive objects.

The discovery of gravitational waves

The announcement, on 11 February 2016, that we had detected gravitational waves for the first time caused a sensation among scientists worldwide. Gravitational waves will allow us to explore fundamental physics, examine the weirdest objects in the universe and possibly even peer back to the universe's earliest moments. The signal was picked up by LIGO's two observatories in Hanford, Washington, and Livingston, Louisiana, on 14 September 2015 (see Figure 4.1). It was created as two black holes circled each other, coming closer and closer until they finally merged into one. The waves were within the frequency range of human hearing. The cataclysmic collision sounds like a thud – or, speeded up somewhat, a chirp.

This sound exactly matches what general relativity predicts. Measuring how the waves' frequency and volume rise and fall, physicists could work out the sizes of the black holes involved: about 36 and 29 times the mass of the Sun. They also worked out that the final black hole was about three solar masses light – all the missing energy being emitted in gravitational waves, which tells us how loud the event was at its source. And comparing

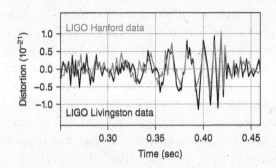

FIGURE 4.1 Huge news from one tiny blip: the signals from gravitational waves picked up by the LIGO observatories in 2015

how powerful it really was to the faint vibrations LIGO detected tells us how far away it occurred – about 1.3 billion light years.

This immediately resolved one open question, as the very existence of such black hole binaries had been contested. Because they are dark, black holes are almost impossible to spot unless something bright – like a star – orbits them.

The masses of the holes in that first merger have puzzled astronomers – they expected to see black holes form when a giant star's core collapses. That should not create anything with more than about 20 times the mass of our Sun.

A second merger was detected in December 2015. These holes were a bit lighter – around 14 and seven solar masses, so within the range expected for collapsed stars.

Death spiral

As well as finding more black hole mergers, LIGO is looking for waves from the death spiral of two neutron stars. Black holes hide their mass behind their event horizons even as they crash, but colliding neutron stars strew hot matter across space, which could help us explore other mysteries. Studying these explosions may explain short gamma-ray bursts – mysterious and incredibly bright electromagnetic phenomena. They might also help explain where much of the universe's heavy elements, like uranium, thorium and gold, are forged. Within the next two years, LIGO should be sensitive enough to detect gravitational waves from any neutron star mergers that happen within the nearest 300,000 galaxies. That means we should see about one signal per month.

Single-event detections are just the start. Put several together and we should be able to gain new insights into the history and composition of the universe as a whole. The signals from several black hole mergers, for example, could be combined to help us understand the nature of dark energy, which is causing the universe's expansion to accelerate. LIGO and other detectors can measure the distance to each merger and, combined with observations from standard telescopes, this might tell us how space has expanded during the time the waves took to reach us, providing a measure of dark energy's effect on space.

Other researchers are hoping to use gravitational wave signals to put relativity to even more stringent tests. For example, they could show whether gravity behaves as relativity predicts over large distances.

The first evidence for gravitational waves

In 1974 astronomers Russell Hulse and Joseph Taylor detected a binary pulsar, a pair of dead stars emitting pulses of radio waves. Hulse and Taylor realized that the two pulsars were losing energy and slowly spiralling towards each other in a way that was exactly consistent with Einstein's equations of general relativity: their orbital energy is being emitted as gravitational waves. The finding earned the pair the Nobel Prize in Physics in 1993.

LIGO's success could see an explosion in gravitational wave detection. India, for example, has long been slated to host a third LIGO detector. Other types of detector could come on the scene, too. The European Space Agency is about to start tests of equipment for the Evolved Laser Interferometer Antenna (eLISA), a huge space-based detector. eLISA should be sensitive to much lower-frequency waves and it might detect merging supermassive black holes, millions to billions of times the mass of the Sun, on the other side of the universe. A preparatory experiment, LISA Pathfinder, is in orbit and began tests in November 2016.

Further ahead, we might see a very different kind of gravitational wave detection. Primordial gravitational waves from the very young universe could be revealed in the cosmic microwave background (see 'The first split second' in Chapter 5), pointing the way to a grand unified theory of nature.

How we tuned into space-time

It took decades of work to prove that gravitational waves are real. Here is the story of how it was done – and the remarkable tool that did it, LIGO (see Figure 4.2).

RAINER WEISS

In 1969 Rainer Weiss (1932–) was a young MIT professor. At the time, gravitational waves were a theoretical curiosity: Einstein himself took years to be convinced by his own prediction that moving cosmic bodies would send out ripples through space-time. Then physicist Joseph Weber (1919–2000) claimed to have recorded one on a xylophone-like instrument he called a resonant bar detector. Weiss takes up the story:

'The students on my course were fascinated by the idea that gravitational waves might exist. I didn't know much about them at all, and for the life of me I could not understand how a bar interacts with a gravitational wave.

'I kept thinking, well, there's one way I can explain how gravitational waves interact with matter. Suppose you sent a light pulse between two masses. Then you do the same when there's a gravitational wave. Lo and behold, you see that the time it takes light to go from one mass to the other changes because of the wave. If the wave is getting bigger, it causes the time to grow a little bit. If the wave is trying to contract, it reduces it a little bit. So, you can see this oscillation in time on the clock.

'For about three months, I thought about how you might do this. First, I thought you couldn't get clocks good enough. But we did some experiments and I learned you could do unbelievably exquisite measurements with lasers. I wrote this up but I didn't publish it. The people at MIT wanted to know how I'd been spending my time, so I put it into the quarterly progress report for my lab. I came to the conclusion that, if you built this thing big enough, you could probably detect gravitational waves.'

BARRY BARISH

Based on Weiss's idea, the US National Science Foundation (NSF) eventually began funding the development of what became the Laser Interferometer Gravitational-wave Observatory (LIGO) in 1979. But progress was slow, and when Caltech physicist Barry Barish (1936–) took the lead on the project in 1994, the NSF had lost confidence in it:

'There was a lot of resistance. It represented the extreme of what you might call a high-risk, high-pay-off project. So we revamped it entirely. Over a period of six months we made it look like a new project. And it was hard because if you had asked me then, could we build what we now know we need to detect gravitational waves, the answer would have been no.

'So the idea, and the words that I used, was that with the initial version of LIGO it would be possible to detect gravitational waves. And it would evolve into a detector, which we called Advanced LIGO, where it would be probable. But in all honesty, we had nothing more than ideas on how to do the Advanced LIGO part. To me, the miracle in this whole thing is that we somehow got financial support for 22 years until we succeeded.'

Michael Landry

LIGO's detectors occupy two sites in the US, in Livingston, Louisiana, and Hanford, Washington. Since joining the project in 2000, Hanford's lead detection scientist Michael Landry has been ensuring that the instrument is as sensitive as possible to the minuscule signals it was built to detect:

> 'Space is a stiff medium, so it doesn't want to vibrate. The detector has to register changes that are about a thousandth the size of a proton. If you were trying to measure the distance between here and our nearest star, Proxima Centauri, it would be like watching that distance change by the width of a human hair.
>
> 'It's an ongoing battle to suppress noise in the instrument. There are ground noises like earthquakes, but a less obvious example is the Earth's ringing – at low frequencies it rings like a bell because of ocean waves hitting against the continental shelf.
>
> 'If you get storms off the coast of Alaska or the Gulf of Mexico, ground motions increase. We have to suppress that motion by registering it with seismometers and feeding it into seismic suppression systems – kind of the way noise-cancelling headphones work by sampling the ambient noise and then playing it with the right phase to cancel it out at your ear.
>
> 'There are also a lot of internal noises to suppress. Things like electronic noise, or quantum noise in the laser. All of that means that the LIGO detectors are the most quiet and the most sensitive detectors ever built.'

In September 2015, just a few days after Advanced LIGO finally came on stream, Michael Landry received notice of an unexplained signal. At first he was convinced it was an 'injection' – an artificial pulse used occasionally to test the instrument:

> 'On the morning of 14 September, I opened my computer and saw an email indicating this event seen in the data literally tens of minutes earlier. I thought, it's probably an injection. It was

so early in the observation phase. It wasn't until later, in the lab, that we determined there was no such injection. We did a whole lot more investigation and it took months to validate it. But it was immediately obvious that if this thing wasn't an injection, it was the best thing we had ever seen.'

The gravitational wave apparently came from the collision of two black holes 1.3 billion light years away. For Rainer Weiss, now a professor emeritus at MIT, it was a long-sought vindication:

'The discovery was spectacular. It was the thing that I would have wanted most, to see the collision of two black holes. If you want to ask what was the reason for building this thing in the first place, it's to check up on whether Einstein's theory works in strong gravitational fields. That was the one place where general relativity had not been tested. And here, suddenly, we have in our hands a thing that says Einstein's field equation, the whole thing, is absolutely right.'

NERGIS MAVALVALA

For Nergis Mavalvala, an MIT physicist who has worked on LIGO for 25 years, this detection is just the beginning:

'One of the amazing things about general relativity is you solve the equations – although that's taken decades – and create templates for how signals should look. Nature was kind in that the very first signal we saw was so clear. Many people expected we would see really weak signals, barely poking above the noise, and that there would be a lot of discussion about whether it was a detection or not.

'The discovery drives us harder because we know there is stuff out there waiting to be observed. You might imagine we would think, "OK, now we've seen it we can pack up and go home." But in fact it's just the opposite. We've seen the very first gravitational wave, but we have so much more to discover. We have a lot to learn about black holes, and then there are neutron stars. Personally, my hope is that we will see something that really has us scratching our heads. Maybe we will have discovered some new object that I can't begin to describe or name.'

Exclusive interview: Einstein reacts to the discovery of gravitational waves

Einstein's theory of relativity predicted gravitational waves. With their discovery in 2016, New Scientist secured an exclusive interview* with the great man himself.

Researchers have discovered the first signs of gravitational waves from two black holes merging, confirming the last prediction of your theory of relativity – exciting stuff, right?

If you ask me whether there are gravitational waves or not, I must answer that I do not know. But it is a highly interesting problem.[1]

That's the thing – they've just found them! What's your reaction?

Theory finds the justification for its existence in the fact that it correlates a large number of single observations, and it is just here that the 'truth' of the theory lies.

What was life like for you when you studied gravitational waves in the 1910s?

Scientifically, I'm having a little breather. I examined gravitational waves, most recently the quantum theory of light emission and absorption, and the causes of lift in flight.

Is it true that you actually found an error in your original 1916 paper on relativity that dealt with

gravitational waves, and had to revisit the matter in 1918?

The important question of how gravitational fields propagate was treated by me in an academy paper one-and-a-half years ago. However, I have to return to the subject matter since my former presentation is not sufficiently transparent and, furthermore, is marred by a regrettable error in calculation.

We all make mistakes. Black holes are also a consequence of relativity, yet few people believed they existed when you were alive. How would you have searched for gravitational waves in your day?

Even the dynamic gravitational fields generated by the rotation of the Earth and the Sun, for which we have the Moon or the inner planets ... as sensitive indicators, remain below the limit of observation.

Not much hope, then. So tell me, what exactly are these gravitational wav es?

I'll send you the wave paper; it's quite pretty.

Thanks, but I'm no Einstein – could you explain it for me without the equations?

I do this the more gladly as there is a certain danger that the – unfortunately – rather complicated mathematical form of the theory threatens to overshadow its simple (and natural) physical content. It is well known that the approximate method of integration of the gravitational equations of the general relativity theory leads to the existence of gravitational waves.[2]

Hmm, I think I need to go back to basics.

I'm sending you a copy of the exposition of the general theory of relativity without the presumption of your really reading it.

Thanks, I guess. So with the last prediction of relativity in the bag, what's next?

It seems that a more complete quantum theory would also have to bring about a modification of the theory of gravitation.

*Quotes are from *The Collected Papers of Albert Einstein*, apart from [1], which is from *Quest: An Autobiography* (1941) by Leopold Infeld, who worked with Einstein, and [2], which is from 'On gravitational waves', a paper Einstein wrote in 1937 with Nathan Rosen.

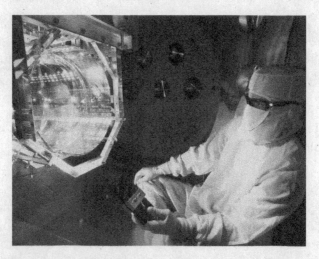

FIGURE 4.2 LIGO, the tool that detected gravitational waves

The next wave

Gravitational wave astronomy has only just begun. By 2021 LIGO itself is due to be upgraded to become a thousand times as sensitive as its 2016 incarnation. The aim is to measure changes in distance as small as 10^{-21} metres – one-ten-thousandth the width of a proton.

Gravitational wave hunters are hoping to detect black holes across the universe, but there are several hurdles to overcome on the way, not least some fundamental laws of physics. LIGO's twin detectors are L-shaped tunnels 4 kilometres long. To detect the stretching and squeezing of space-time caused by a passing gravitational wave, physicists shine a laser down each arm to reflect off a mirror at the end. When the beam returns to the detector's elbow, they recombine it with the light from the other arm to see whether the beams are still in phase, meaning they went the same distance. If not, they've caught a gravitational wave.

To be sure of this, every other wobble must be accounted for. That includes anything that can move the mirrors: waves crashing on the shore, the rumble of a passing car, even the laser itself. One strategy is to keep the mirrors away from the ground: the team suspend them from an isolated scaffold. They also measure the ground's motion with seismometers and nudge the mirrors in real time to cancel it out.

But the seismometers cannot tell the difference between the mirrors shaking as the result of an earthquake (which they can pick up from as far away as Australia) and other effects that move the mirrors. A strong wind can be enough to tilt the building where the seismometers are kept, making them move the mirrors when they shouldn't. So the team are working on suspending the seismometers on thin strands of glass to isolate them from non-earthquake buffeting.

A more fundamental limit is imposed by quantum mechanics. At wave frequencies higher than about 1 kilohertz, Heisenberg's uncertainty principle becomes a factor. This says that, for certain pairs of properties, the more accurately you can measure one the less accurately you can measure the other. For LIGO, the two relevant properties are the brightness and phase of the light waves.

Fortunately, you can send the light through a special kind of crystal to squeeze it into preferring one property, which can be measured with great confidence at the cost of higher uncertainty for the other. In this case, the phase detection is boosted by reducing the brightness and therefore the number of photons the experiment picks up. This is already done in LIGO, but a new way to squeeze light using specialized micrometre-scale mirrors will be added in the upgrade.

Soon LIGO will be joined by more detectors, including VIRGO in Europe and KAGRA in Japan. As well as confirming observations, these instruments will enable astronomers to triangulate the position of black hole mergers and other events, to help optical and other telescopes look for their sources.

Upgraded LIGO and its partners could capture new kinds of gravitational wave sources, such as starquakes on neutron stars. But to see these in more detail and capture more events from a wider swath of the universe, we will need an even more sensitive detector. A group in Germany is working on the Einstein telescope, which will have 10-kilometre-long arms and be located underground for greater sensitivity. And physicists are already dreaming of a detector with arms 40 kilometres long. Tentatively called the Cosmic Explorer, it would be sensitive to gravitational waves of lower frequency arriving from much greater distances – potentially all the way back to the collapsed remnants of the very first stars.

Are there particles of gravity in these waves?

About 150 years ago the Scottish physicist James Clerk Maxwell devised equations that predicted the existence of electromagnetic waves moving at the speed of light. That was what physicists today call a classical field theory. It works very well for long-wavelength radiation such as radio waves. It was only through considering short-wavelength, high-frequency radiation – such as visible light, ultraviolet and X-rays – that a quantum description emerged in the early twentieth century, leading to the notion of tiny particles of light called photons.

General relativity, Einstein's theory that predicts gravitational waves, is a classical field theory like Maxwell's. Just as we can describe radio in terms of waves – without worrying about the photons that make them up – the gravitational waves that we have detected are of long enough wavelength that we can indeed deal with them as smooth waves.

In the future, we hope to be able to detect gravitational radiation with shorter wavelengths, where the wave-like description starts to break down. Then we would need to consider it in terms of particles of gravity, gravitons. If so, and if the waves are at some level made of particles, then those particles would have to be massless, or nearly so. In general relativity, gravitational waves travel at the speed of light, which is only possible for massless particles.

A future theory of quantum gravity might have gravitons with a very small mass; then gravitational waves would move a little more slowly than light. Our results can already be used to put an upper limit on the graviton's mass, because a very massive graviton would affect the shape of the waves predicted by general relativity for the merger of two black holes.

Can we make anti-gravity devices?

Although no one has yet managed to do it, the notion of building a gravity shield has a long history. Perhaps the most famous attempt was by Russian émigré scientist Evgeny Podkletnov. In 1992 Podkletnov published a paper in which he claimed to have detected a 2 per cent weight reduction around a spinning disc made out of a ceramic superconductor.

Martin Tajmar, a researcher at Austrian Research Centres, published a similar claim in 2003 and was able to pursue the research further with funding from the European Space Agency. Three years later Tajmar and the ESA announced that they had measured an effect in a spinning superconductor that might, with further development, be harnessed to affect gravity. Others have not been able to replicate this effect, but relativity does not rule out the possibility that the bent space-time that gives rise to gravity's pull can be 'unbent'.

With the right arrangements of matter and energy, it should be possible to diminish or enhance the influence of gravity. For example, you could use an effect called gravito-magnetism. According to general relativity, the mass of a rotating body drags space-time around with it like a whirlpool. Unfortunately, this effect is extremely small in practice – and it is not clear that a spinning superconductor has any gravito-magnetic influence. But it is not entirely out of the question that, one day, someone will find a way to harness repulsive gravitational effects for propulsion or gravity shielding.

5
Into the cosmos

Cosmologists study the universe as a whole – its birth and growth, its size and shape, and its eventual fate – and they try to understand it by using mathematical models based on general relativity.

A primordial primer

The vast scale of our universe became clear in the 1920s, when Edwin Hubble proved that 'spiral nebulae' are actually other galaxies like ours, millions to billions of light years away. He also found that their light is reddened, which can be explained if they are receding from us. The universe is expanding.

This led to the Big Bang Theory. If everything is flying apart now, it was once presumably packed much more closely together, meaning that the newborn universe was dense and hot. A rival idea, the Steady State Theory, holds that new matter is constantly being created to fill the gaps generated by expansion; but the Big Bang triumphed in 1965 when Arno Penzias (1933–) and Robert Woodrow Wilson (1936–) discovered cosmic microwave background radiation. This is relic heat radiation emitted by hot matter in the early universe, 380,000 years after the first instant of the Big Bang.

The growth of the universe can be modelled using relativity: if you assume that on the largest scales the universe is uniform, then the complexities of the theory can be reduced to fairly simple equations called Friedmann models, which describe how space expands and evolves.

According to these models, the overall shape of space-time on the scale of the universe is free to be curved, either inwards like the surface of a sphere or flared out like a saddle. Instead, observations suggest that it is poised between the two kinds of curvature, and almost exactly flat. This is not prohibited by the Friedmann models, but it seems strangely fine-tuned. One explanation is the theory of inflation. This states that during the first split second of existence, space expanded at terrifying speed, flattening out any original curvature. Today's observable universe grew from a microscopic patch of the original fireball. This would also explain the horizon problem – why it is that one side of the universe is almost the same density and temperature as the other.

Dents in space–time

The universe is not totally flat, of course. Galaxies put little local dents in space–time. And in 1990 the COBE satellite detected ripples in the cosmic microwave background, the signature of primordial density fluctuations. These slight ripples in the early universe may have been generated by random quantum fluctuations in the energy field that drove inflation. Gravity amplifies the original fluctuations, pulling denser patches of matter together to become stars, galaxies and galaxy clusters. Today, galaxies are scattered throughout the universe in a foamy pattern, in knots, strings and walls that enclose bubble–like voids. These structures have scales of up to about a billion light years.

Visible matter does not have enough gravity to create the amount of structure we see: it has to be helped out by some form of dark matter. More evidence for this dark stuff comes from galaxies that are rotating too fast to hold together without extra gravitational glue.

Dark matter cannot be made from protons, neutrons and electrons. When the universe was less than three minutes old, protons and neutrons fused to make deuterium (a heavy form of hydrogen) and traces of other light elements. Cosmologists calculate that, if there had been much more ordinary matter than we see, the dense cauldron would have brewed up a lot more deuterium than is observed. So dark matter must be something exotic, probably generated in the hot early moments of the Big Bang. Physicists have thought up plenty of options, including hypothetical particles such as WIMPs (weakly interacting massive particles) and lighter axions. Primordial black holes, forged in the Big Bang, could also form dark matter.

Another dark mystery emerged in the 1990s, when astronomers found that distant supernovae are surprisingly faint – suggesting that the expansion of the universe is not slowing down as everyone expected, but accelerating. The universe seems to be dominated by some repulsive force, which has been called dark energy. It may be a cosmological constant (vacuum energy) or a changing energy field such as quintessence – or gravity may behave differently at long ranges, becoming repulsive.

The microwave background

The WMAP and Planck space missions put this standard picture of cosmology on a firm footing by precisely measuring the spectrum of fluctuations in the microwave background. The data fit a universe 13.8 billion years old, containing about 5 per cent ordinary matter, 26 per cent dark matter and 69 per cent dark energy.

Many big questions remain. We do not know the true size of the universe, or even whether it is infinite. Nor do we know its topology – whether space wraps around on itself. We do not know what caused inflation, or whether it has created a plethora of parallel universes far from our own, as many inflationary theories imply.

Another mystery is why the universe favours matter over antimatter. Early in the Big Bang, when particles were being created, there must have been a strong bias towards matter, otherwise matter and antimatter would have annihilated each other and there would be almost nothing left but radiation. The standard model of particle physics does not explain this material bias.

The true beginning, if there was one, is still unknown, because in relativistic models the Big Bang begins at a singularity, where the equations break down. The final fate of the universe is also veiled, as it depends on the nature of dark energy and how that strange stuff behaves in the future.

Galaxies might become isolated by acceleration, or all matter could be destroyed in a big rip, or the universe might collapse in a big crunch – perhaps re-expanding as a cyclic universe. To understand both the beginning and the end of the universe, we will probably need a quantum theory of gravity.

The Big Bang begins

More than 50 years ago, the universe's genesis story was confirmed – by accident. The Canadian-American cosmologist Jim Peebles (1935–) recalls the struggle to convince doubters.

'I was there at the birth of the Big Bang – in a manner of speaking. In May 1964 Arno Penzias and Robert Wilson had just recorded their first measurements of astronomical microwave radiation, from the supernova remnant Cassiopeia A. Based at the Bell Telephone Laboratories in Holmdel, New Jersey, they were using a horn antenna that had been assembled to study microwave communication, an early step towards today's cellphone technology. But Penzias and Wilson found a problem that had already been bothering Bell's engineers. There seemed to be more microwaves coming from the sky than anyone had expected.

'I was a young theorist just down the road at Princeton University, in the research group of Bob Dicke. Bob was a fan of the idea that the universe began in a hot, dense state, and he was exploring the idea that it should have left behind a sea of radiation spread across the sky. Bob had set two members of his group, Peter Roll and David Wilkinson, to building a receiver to catch this radiation, and suggested I look into the theoretical consequences of detecting or not detecting it.

'The following February I presented our idea in a colloquium, and a few weeks later Bob received a phone call from Penzias. When Bob, Peter and David visited Holmdel, they discovered that we had been scooped. I don't remember any expression of regret by Bob or any of us. Instead, there was excitement that something was there to be measured and analysed.

'The sea of noise that was so troubling to Penzias and Wilson is the cosmic microwave background (CMB), which is now known to be clinching evidence for a big bang. But it wasn't immediately seen that way, as a glance back through the pages of *New Scientist* shows. In 1976 Martin Rees, then recently elected to the illustrious Plumian Professorship of Astronomy at the University of Cambridge, writes that "no plausible alternative theory" could account for the CMB's observed characteristics (2 December 1976, page 512). Yet five years later, Rees's Plumian predecessor, Fred Hoyle, still sees a chance for the rival steady state theory, writing that the latest CMB measurements "differ by so much from what theory would suggest as to kill the big-bang cosmologies" (19 November 1981, page 521). The steady state theory, which Hoyle co-devised in 1948, suggests that matter is continually created in the expanding universe, with new galaxies forming to fill the spaces that open up as already existing ones move apart. In this picture, the universe's past was no hotter or denser than its present.

'A sure sign of our universe's origin in a hot, dense state would be a characteristic spectrum of intensities at different wavelengths known as a thermal Planck spectrum. Roll and Wilkinson's experiment soon added another data point, and Wilkinson eventually added many more,

FIGURE 5.1 The cosmic microwave background (CMB), the oldest light in the universe, as seen by the Planck space probe

culminating in his leading role in the COBE satellite mission. In the early 1990s COBE finally showed that the CMB spectrum is close to a Planck form with a temperature of around 2.73 kelvin. Few, apart from Hoyle and his close associates, doubted the universe's origin in a big bang by then.

'COBE and its follow-up missions – NASA's WMAP, the European Space Agency's Planck probe and a host of others – have given us measurements of how the CMB departs from an exactly uniform sea of radiation (see Figure 5.1). These measurements inform us about the history of expansion of the universe and the nature of its material contents. We now have an excellent fit between cosmological theory and measurements, albeit one that requires two hypothetical components: unseen cold dark matter to keep galaxies clumped together, and the cosmological constant, needed to account for the accelerated expansion of the universe revealed by measurements of far-off supernovae.

'Today, ever more detailed explorations of the CMB could be taking us back to a universe even beyond general relativity (see 'The first split second' later in this chapter). But a crucial way station was reached 50 years ago when an annoying hiss in a glorified telecommunications antenna told a story of how the universe began.'

Intruders from another universe

On a large scale the cosmos should be plain, but it's not. Windows into other dimensions could explain mysterious objects billions of light years across.

As our observations of the cosmos come into sharper focus, astronomers are beginning to identify structures bigger than any seen before:

- A giant hole in the cosmic web of galaxies
- A string of quasars billions of light years across
- A ring of explosions that spans a fair slice of the visible universe.

Because these megastructures sit uncomfortably with mainstream cosmology, one researcher has suggested that they are illusions projected from another dimension: the first tantalizing evidence of realities beyond our own.

Ever since Copernicus proposed that the Earth's place among the stars is nothing special, astronomers have regarded it as fundamental idea. It evolved into the cosmological principle: that nowhere in the universe is special. There are patches of individuality on the level of galaxies, clusters and even superclusters of galaxies, of course, but zoom out far enough and the universe should exhibit a drab homogeneity.

The cosmological principle is just an assumption, but on the whole it seems to hold up. The latest data show that the cosmological principle applies on scales of roughly a billion light years, with the average amount of material in any given volume more or less the same – but not everywhere. Take that giant hole in the universe: a void almost 2 billion light years wide. There are 10,000 fewer galaxies in that part of the sky compared with the universal average, says co-discoverer András Kovács of the Institute for High Energy Physics in Barcelona, Spain. A big empty patch almost double the size of the cut-off stands out like a sore thumb. Kovács and his team call this vast expanse a supervoid, and believe that another like it might explain another colossal anomaly: a giant cold spot in the cosmic microwave background that has been puzzling astronomers for more than a decade.

Could ancient life have emerged in the Big Bang's glow?

The cosmic microwave background is too faint and frigid today to incubate life, which as far as we know needs the warmth of a star or at least a hydrothermal vent. But around 15 million years after the Big Bang, the CMB glow would have been warm enough to make the whole universe one large life-friendly zone. This epoch would have lasted a few million years, perhaps enough time for microbes to emerge but probably not complex life.

The infant universe

Our colour maps of the CMB are littered with red and blue speckles representing the slightly hotter and cooler regions of the infant universe. In 2004 the WMAP saw one cold spot much larger than the others; then the European Space

Agency's Planck satellite observed it too. If a supervoid sits in the same direction, then CMB photons originating from beyond the supervoid would have had to pass through it. Thanks to the accelerating expansion of the universe, the photons emerging from this barren area would find that matter was less densely packed than when they went in, leading to a drop in the gravitational potential they experienced and, consequently, their energy – effectively cooling them down.

The supervoid isn't the half of it. As far back as 2012, a team led by Roger Clowes at the University of Central Lancashire, UK, claimed to have found an enormous structure strung out over 4 billion light years, more than twice the size of the supervoid. This time it wasn't an empty patch of space but a crowded one. Known as the Huge Large Quasar Group, it contains 73 quasars – the bright, active central regions of some galaxies. Astronomers have known since the early 1980s that quasars tend to huddle together, but never before had a grouping been found on such a large scale.

Then, in 2015, a team of Hungarian astronomers uncovered a colossal group of gamma-ray bursts (GRBs) – highly energetic, short-lived flashes of energy erupting from distant galaxies. These GRBs appear to form a ring a massive 5.6 billion light years across, 6 per cent of the size of the visible universe.

Such apparent violations of the cosmological principle make astronomers deeply uncomfortable. They may turn out to be mistakes. For example, one study from 2013 calculates that there is a fairly high probability of seeing an apparent structure within what is really a random distribution of quasars – although Clowes's group disputes that conclusion.

Rainer Dick, a theoretical physicist at the University of Saskatchewan, Canada, believes that attempts to brush these cosmic megastructures aside are misguided. He says they should be embraced as our best bet of keeping the cosmological principle alive. All we have to do is accept that they don't actually exist; instead, they are the first evidence of other dimensions intruding into our own, leaving dirty footprints behind on our otherwise smooth and homogeneous cosmic background.

It seems an audacious proposal, but it builds on a solid foundation of theoretical work. Conjuring up other dimensions beyond our own is nothing new. For decades, many theorists have regarded the existence of extra dimensions as our best hope of reconciling Einstein's general relativity with that other bastion of twentieth-century physics, quantum theory. A marriage between these two seemingly disparate concepts, one dealing with the very large and the other with the very small, would yield what is often called a theory of everything, a one-size-fits-all framework capable of describing the universe in its entirety.

M-theory

One popular candidate is M-theory, an extension of string theory that famously suggests we live in an 11-dimensional universe, with the other seven dimensions curled up so tightly as to drop out of sight. It's an elegant and mathematically appealing framework, with one major failing: the lack of solid predictions offering opportunities to verify it. Dick's work on a generalization of string theory, known as brane theory, might provide just such a prediction, and resolve the cosmological principle dilemma at the same time. In brane theory, our universe is a four-dimensional membrane floating in a sea of similar branes spanning many extra dimensions. The theory suggests that we might just be able to spot the effects of a neighbouring brane overlapping with ours.

To measure our distance to far-off objects, astronomers exploit an effect known as redshift (see also Chapter 1). Any object moving away from us because of the universe's expansion will have its light stretched out to longer wavelengths, and tell-tale lines in the spectrum will appear shifted towards the red end of the spectrum. The farther away the object, the more the lines will shift. If astronomers see many objects all exhibiting the same redshift, they will interpret that as some form of structure, just like the GRB ring or the huge quasar group (Figure 5.2A).

However, another brane overlapping with our own might skew these redshift measurements. Photons in one brane would exert a force on charged particles in another – a phenomenon Dick calls 'brane crosstalk'. That could change energy levels within atoms, shifting the spectral lines of light they absorb or emit. In other words, brane crosstalk would create a redshift that has nothing to do with the expansion of the universe. It would produce an apparent pile-up of objects at one redshift and a distinct lack of objects at another – an illusion that would make a homogeneous universe appear to contain massive structures and enormous voids (Figure 5.2B).

Of course, it's hardly an open-and-shut case. Moataz Emam, from the State University of New York College at Cortland, warns that some of the assumptions in Dick's theory have been criticized in the past. But the model is certainly testable: Emam suggests observing parts of the sky where high-density regions sit next to apparent barren patches. If the discrepancy in redshift measurements is identical in all cases, it might well suggest that our brane is overlapping with another.

With the help of the Sloan Digital Sky Survey (SDSS) – the most detailed three-dimensional map of the universe ever made – Dick is planning to scour the databases for redshift data that could support his theory. His quest to cut the universe's largest objects down to size might lead to new monsters arising in their place. The discovery of branes beyond our own would make a nonsense of our concept of cosmic homogeneity. In a vast multiverse of interacting membranes, the cosmological principle might not be worth saving after all.

Atoms emit light at
characteristic wavelengths.

Light from far-off cosmic sources
is stretched to longer, **redder
wavelengths** by the universe's
expansion. The extent of
the stretching tells
us our distance
to the object.

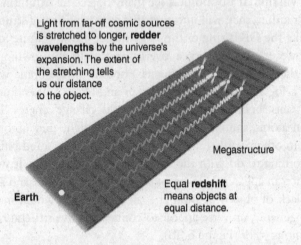

Megastructure

Earth

Equal **redshift**
means objects at
equal distance.

FIGURE 5.2A The traditional explanation of megastructures

The influence of extra dimensions
could stretch the natural wavelength
of light emitted.

**Overlapping
dimensions**

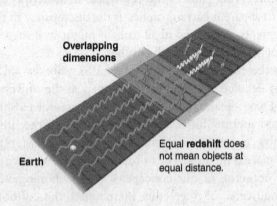

Earth

Equal **redshift** does
not mean objects at
equal distance.

FIGURE 5.2B An alternative explanation of megastructures

The first split second

The secrets of the Big Bang hide behind an impenetrable wall of fire, but there is a way to look back further than we once thought possible, to a time before stars and galaxies, before atoms and molecules, before even protons and neutrons existed, to the origin of everything, the detonation of the explosion that became our universe. All you need is a pair of sunglasses.

Just looking at the stars means seeing into the past, because light takes time to travel from a distant star to our eyes. If you look far enough out into space, you might think you could reach back to the birth of the universe, but a hot, opaque soup of electrons and atomic nuclei in the early universe makes this impossible. Only some 380,000 years after the beginning did it cool down enough to become transparent. However powerful your telescope, you can't see anything that happened before that time.

Even so, physicists would love to look back to a tiny fraction of a second after the beginning, into a hypothetical, almost mythical era known as inflation. This has been devised to patch up some problems with the otherwise successful theory of the Big Bang. According to Einstein's theory of general relativity, matter and energy can bend space-time – but on large scales our universe turns out to be flat. Also, there is nothing in the model to say that widely separated regions should look similar, yet galaxies cluster in the same numbers and patterns on one side of the universe as on another. Finally, there must have been some small density fluctuations in the early universe to gradually gather matter together by their gravity, or else there would be no galaxies today – just a smooth, tenuous gas filling space. But, again, the basic Big Bang theory doesn't explain why these fluctuations occurred.

Inflation

To solve these puzzling problems, inflation was proposed in 1980 by Alan Guth, then at the Massachusetts Institute of Technology. He assumed that a brief burst of acceleration took space and stretched it out in all directions. What we now call the observable universe would have started at subatomic size and blown up to a few centimetres across in a fraction of a second. Any initial curvature would have been flattened by this expansion, and any differences in temperature or density, say, would have evened out. Tiny quantum fluctuations in the energy fields filling space would have been magnified to form those early density fluctuations (see Figure 5.3).

There are many theories of inflation. All agree that some pervasive energy field blew space apart and then vanished, leaving behind a sea of subatomic particles; but they do not agree on what that initial field actually was. Theories range from the weird to the even weirder, based on different ideas of how the forces of nature behave at high energies.

It might seem impossible to find out what was really going on in that first instant, given the opaque nature of the early universe. But it was not opaque to everything. The sudden end of inflation should have sent shudders through space-time known as gravitational waves. These waves would have passed through the primordial fireball and left their mark upon it as they squeezed and stretched space-time.

We can still see the glow of that fireball, stretched and diluted by the expansion of space into what is known as the cosmic microwave background (CMB). Every microwave photon vibrates in some direction, called its plane of polarization. Usually, each photon is polarized in a different, random direction, but various processes can imprint a large-scale pattern on polarization. Gravitational waves should make distinctive swirl-like patterns.

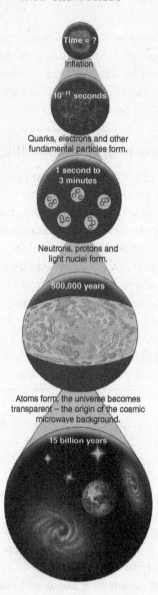

Time = ?

Inflation

10^{-11} seconds

Quarks, electrons and other
fundamental particles form.

1 second to
3 minutes

Neutrons, protons and
light nuclei form.

500,000 years

Atoms form, the universe becomes
transparent – the origin of the cosmic
microwave background.

15 billion years

FIGURE 5.3 After the Big Bang

To see this, we need the microwave equivalent of Polaroid sunglasses: a set of parallel wires that stop waves with one polarization while letting others pass. By rotating this set-up and measuring how much radiation leaks through, you can work out the direction of polarization for every spot on the sky.

Was there anything before the Big Bang?

Inflation erased any record of what came before the rapid expansion, so we are probably unable to answer this question by looking at the sky. In addition, the universe started out so hot and dense that the equations we use to describe its expansion and evolution break down.

A theory that unites these equations with quantum mechanics could make mathematical predictions of what, if anything, existed before the Big Bang, but such a theory remains elusive. Of course, that has not prevented people from speculating about bouncing universes that cycle through endless bangs and crunches, or multiverses that bud off from a single original universe.

Several experiments have tried to sift out the gravitational wave imprint. In 2014 researchers at the BICEP2 detector claimed to have found it – but, in fact, their observation was caused by interstellar dust. In the high, dry places of the world, scientists are racing to spot the signal using new telescopes and detectors (see below).

If they succeed, physicists could work out exactly when, how and why inflation happened. In most theories, inflation is triggered some time between 10^{-36} and 10^{-33} seconds from time zero, when an original cosmic superforce split into three independent forces: the strong and weak nuclear forces, and electromagnetism.

This split is a phase transition, like the transition from liquid water to ice, and it releases energy to drive inflation. The gravitational wave pattern could give us insights into how the superforce works – bringing us to a new 'grand unified theory' of physics that describes the true fundamental structure of matter. It might reveal whether there are extra, hidden dimensions to space, which would delay inflation by perhaps as much as 10^{-14} seconds.

Then again, we might see a pattern that can't be explained by inflation at all. The theorists would have to start from scratch.

Detectors and telescopes

The most inhospitable places on Earth are perfect for spying on the first moments of the universe's existence.

One of the biggest telescopes on the planet looks very small against the vast Antarctic landscape. The South Pole Telescope (SPT) was built to study the cosmic microwave background (CMB), giving us a picture of the universe 380,000 years after the Big Bang. Using a camera installed at the telescope, scientists hope to pick up the imprint of primordial gravitational waves, and so shed light on the first trillionth of a trillionth of a trillionth of a second after the birth of the universe (see 'The first split second', above).

The camera is designed to measure polarization. Just as sunlight is polarized when it reflects from a lake or road, the microwave background is polarized as it scatters off electrons on its journey through the universe, and gravitational waves should subtly change the polarization pattern. Spotting this effect will be difficult, like listening for the sound of a cricket during a rock concert, because the weak signal from gravitational waves is on top of a much stronger one from density fluctuations in the early universe.

FIGURE 5.4 The National Science Foundation's 10-metre South Pole
Telescope (SPT)

Water vapour in the air absorbs microwaves, so to see
the CMB you need to be high and dry. The South Pole
is at an altitude of 2,830 metres and the air is extremely
dry. But it is not the highest site. The Array for Microwave Background Anisotropy, AMiBA, is some 3,400
metres up on the slopes of Mauna Loa on Hawaii. Even
higher, at 5,200 metres near the summit of Cerro Toco in
the Atacama desert in Chile, are several more detectors
looking for the gravitational wave signal, including the Polarbear experiment, the ACTPol camera and
CLASS – the Cosmology Large Angular Scale Surveyor.
Other detectors dangle, higher still, from balloons above
Antarctica, Australia and New Mexico.

6
Dark matter

Our cosmos is much more than the galaxies and gas we can see. It is also made of mysterious stuff called dark matter, which far outweighs ordinary matter and acts as a gravitational glue to form stars and galaxies.

Shedding light on dark matter

A few decades ago, we thought we understood the substances that fill the universe, but no longer. We now know that the atoms making up everything visible in the cosmos – from galaxies to planets to clouds of interstellar gas and dust – represent less than 20 per cent of the total matter out there. The remaining 80 per cent is dark matter, invisible to conventional telescopes. But if we cannot see it, how do we know it is there?

We cannot weigh the Sun or a planet directly. Instead, we determine its mass by measuring how its gravitational pull influences the motion of objects around it. In the same way, it is possible to measure the mass of a galaxy, or even a cluster of galaxies, by observing how fast stars or other objects move around it. In 1933 the Swiss astronomer Fritz Zwicky (1898–1974) applied this principle to the Coma Cluster, a group of more than 1,000 galaxies some 300 million light years from us. He found that the individual galaxies were zipping around far too rapidly for their gravity to keep them bound together in a cluster. By rights, they should have been flying off in different directions (see Figure 6.1).

Zwicky's puzzling results drew little attention until the late 1960s, when Vera Rubin (1928–) at the Carnegie Institution in Washington DC measured the Doppler shift of clouds of hydrogen gas in several distant galaxies. This showed that the speeds at which the clouds were orbiting the centre of their galaxies seemed to require far more mass than could be accounted for by visible material.

Without dark matter, the very existence of many apparently stable galaxies would defy the laws of physics. The fact that they do exist remains among the most compelling reasons to think that there must be more to the cosmos than meets the eye.

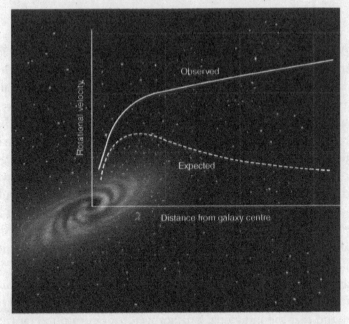

FIGURE 6.1 Evidence for dark matter: stars near the edge of galaxies are travelling too fast to be held in orbit merely by the gravity of the matter we can see in the galactic centre

The cosmic microwave background

Although we still cannot see the stuff itself, we see evidence for dark matter everywhere we look, for example in the radiation known as the cosmic microwave background (CMB). By studying the patterns of slightly hotter and colder patches in the CMB, we have been able to learn a great deal about our universe's history and composition (see Figure 6.2). Among other things, these variations in the CMB tell us how matter was distributed throughout space in the early universe. Because dark matter began clumping under the influence of gravity earlier than normal matter did, its influence can be seen in numerous

Universe 380,000 years old

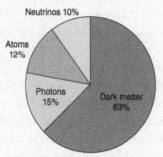

The imprint left on the cosmic
microwave background shows that dark
matter accounted for most of the mass
and energy of the early universe.

Modern universe

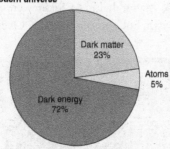

Today, both dark matter and visible matter seem
to be dwarfed by dark energy, an even more
mysterious substance accelerating the
expansion of the universe (see Chapter 7).

FIGURE 6.2 The early and the modern universe compared

small hot and cold patches, each covering an angle in the sky of
0.25 degrees or so.

The pattern of these spots allows us to determine how
much dark matter must be present. It turns out that for every
gram of stuff that we can see in the cosmos there must be 4 or
5 grams we cannot see.

Even if dark matter were not needed to prevent galaxies flying apart, supercomputer simulations suggest that the cosmos would look very different if it did not exist. These simulations track the movement of billions of particles through cosmic time, with the aim of better understanding why the universe has ended up the way it has. When atoms in a gas of ordinary matter are compressed, they collide more frequently. This interaction tends to push the atoms apart and so hinders gravity from compressing the gas further. Dark matter particles, on the other hand, interact with each other only feebly and so clump much more readily. Simulations that embody these properties show that, as the universe expanded and evolved, the first structures to form would have been clumps, or 'halos', of dark matter (see Figure 6.3).

The first dark matter halos to form were probably about as massive as the Earth, but far more diffuse. Over time, they began

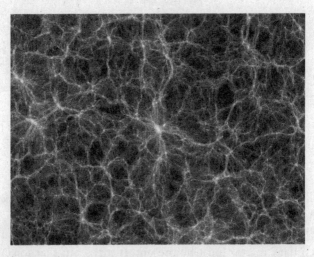

FIGURE 6.3 Dark matter simulation showing the distribution of dark matter in the local universe; the frame is 253 megaparsecs (824 million light years) in distance across

to merge and became steadily larger. Eventually, some became massive enough to attract large quantities of hydrogen, helium and other conventional matter – the seeds of the first stars and galaxies.

The agreement between the shapes and sizes of the structures derived in dark matter simulations and those observed in our universe is striking. It leaves little doubt that dark matter is not only real but also that it formed the nurseries in which galaxies such as our own Milky Way formed.

What is dark matter?

The short answer is that we still don't know. It must be invisible, or at least very faint, so it cannot be made of anything that significantly radiates, reflects or absorbs light. That rules out conventional atom-based matter. We once thought that it might be large objects such as black holes or exotic types of star – neutron stars or white dwarfs – that are nearly invisible to our telescopes. But observations seem to have ruled out these 'massive astrophysical compact halo objects', or MACHOs. The concentrated gravity of a MACHO would deflect passing light on its way to us from distant stars. We see such gravitational lensing effects, but only rarely: enough for MACHOs to account for at most a few per cent of the missing mass.

Most cosmologists now think instead that we are swimming in a sea of dark matter – a gas of weakly interacting massive particles, or WIMPs – that pervades the whole galaxy, including our solar system. However, none of the particles discovered over the past century fits the bill. Dark matter must be something completely new. Dozens of different possibilities have been suggested over the years. Proposals range from heavy, neutrino-like particles, to ultra-light things known as axions, to subtle twists on ordinary matter (see 'Strangely familiar' below), to truly bizarre possibilities such as particles that are moving through extra dimensions of space.

Supersymmetry

To many physicists, however, there is a clear favourite among these possibilities: particles predicted by a class of theories that goes by the name of supersymmetry. In our world, there are two classes of particle: fermions and bosons. Fermions are particles such as electrons, neutrinos and quarks that make up what we normally think of as matter. Bosons are the particles responsible for transmitting the forces of nature. The electromagnetic force is nothing more than bosons – photons, in this case – shuttling back and forth between electrically charged particles.

Supersymmetry postulates that, for each type of fermion, a type of boson with many of the same properties must also exist (see Figure 6.4). The electron, for example, has an as yet undiscovered bosonic partner called a selectron. Similarly, the photon would have a fermion analogue known as a photino.

Among the new particles in supersymmetry is one that is likely to be stable and have the characteristics required of a dark matter candidate. It is the lightest version of a class of particle known as a neutralino. If neutralinos do exist, the lightest version would probably have been produced in the first seconds after the Big Bang in quantities similar to what is needed to account for the dark matter in our universe today.

There is, of course, a catch. No one has seen a supersymmetric particle. Physicists suspect that the superpartner particles, if they exist, are considerably heavier than their ordinary counterparts, making them difficult to create or discover in experiments. But the Large Hadron Collider, housed at the CERN particle physics laboratory near Geneva, Switzerland, has recently reached the kind of energies where we expected supersymmetric particles to pop up. It will not be able to detect them directly, but it could infer their presence from imbalances

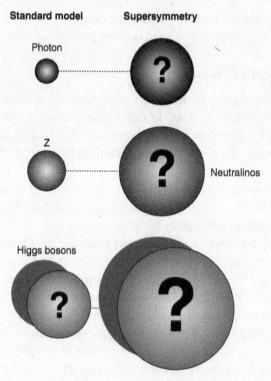

FIGURE 6.4 Standard model versus supersymmetry theories: in supersymmetry theories, neutralinos pop up as heavier partners of the photon, the Z boson that mediates the weak nuclear force and the as yet undiscovered Higgs bosons

in the energy and momentum being carried away from particle collisions. So far, nothing has been discovered.

Other physicists are trying to snare dark matter particles as they sweep through our planet. This is difficult. Dark matter particles must interact only feebly with normal matter in order to be dark, and yet they must be picked up against a very noisy backdrop of other particle interactions from natural radioactivity and cosmic rays. It's like standing on a busy street corner, listening out for the sound of a pin dropping to the ground.

Dark matter hunters

Some dark matter hunters have gone underground, setting up detectors in mines to shield them against cosmic rays. So far, these deep detectors have not seen any dark matter particles. Occasionally there is a hint of a detection, but with further data it turns out to be a mirage.

While some delve underground, others look to space. Here, the aim is to see the energetic particles that can be created when massive dark matter particles interact and annihilate one another. Unconventional telescopes built to detect gamma rays, antimatter and neutrinos have picked up signals tantalizingly similar to those predicted to come from dark matter annihilations. In 2008 the satellite-borne PAMELA experiment discovered that a surprisingly high proportion of the cosmic rays travelling through space were not ordinary matter, but antimatter. That might be down to dark matter annihilations – although it is more likely to come from well-known sources of antimatter in our galaxy, such as the fast-rotating neutron stars known as pulsars. NASA's Fermi Gamma-Ray Space Telescope has also seen a bright source of gamma rays from the centre of our galaxy that looks very much like the signal expected from annihilating dark matter particles. At the moment, this seems more persuasive than the PAMELA results, but it, too, might be caused by some previously unknown population of astrophysical objects.

Could it be that this is all a wild-goose chase, and dark matter does not really exist? In 1983 the Israeli physicist Mordehai Milgrom (1946–) suggested that the higher-than-expected speeds of stars moving around galaxies might be explained another way – if gravity worked differently than predicted by

Newton or Einstein. He pointed out that the observed galactic rotations could be explained if Newton's second law of motion – force equals mass times acceleration, or $F = ma$ – were modified to make the force of gravity proportional to the square of the acceleration at very low accelerations.

In recent years, however, Milgrom's proposal, called MOND, for 'modified Newtonian dynamics' – has suffered some serious setbacks. In particular, it has not managed to explain convincingly the dynamics of galaxies within clusters. Observations in 2006 revealed a pair of merging galaxy clusters, known collectively as the Bullet Cluster, whose motion indicated that their gravity was not centred on the gas and stars, as would be expected according to MOND. This suggests that dark matter has shifted the centre of gravity elsewhere, and most cosmologists longer consider MOND to be a viable alternative to dark matter. The Bullet Cluster does not rule out all modifications of gravity – and some radical alternatives to relativity claim to explain some of dark matter's effects (see Chapter 8) – but it does imply that we need dark matter in any case.

The sensitivity of dark matter detectors has been improving by an impressive factor of 10 approximately every two years, so the first unambiguous evidence for particle dark matter might appear within the next few years. Then we can begin to shed light on dark matter's properties in detail.

Strangely familiar

Dreaming up new particles to explain the universe's missing mass has got us nowhere. Is dark matter just normal stuff in disguise?

In July 2015 an unexpected visitor appeared at CERN's Large Hadron Collider. Named the pentaquark, this peculiar particle represents a fundamentally new way to aggregate the basic building blocks of matter. This sort of thing is music to Glenn Starkman's ears. A theoretical physicist at Case Western Reserve University in Cleveland, Ohio, Starkman proposes a bold idea: that other configurations of ordinary matter are out there, perhaps enough to make the universe's elusive dark matter.

To form the matter that surrounds us, elementary particles come together in certain standard configurations. Quarks combine in threes to form compound particles known as baryons, including the protons and neutrons that make up atomic nuclei. We also know of short-lived combinations of a quark and an antiquark, known as mesons.

But quarks are quirky: they never float around freely, thanks to a peculiar property of the strong nuclear force that binds them. When the distance between quarks is small, the force is weak. But as that distance grows, the force gets stronger and the quarks are pulled back together. The force is complicated in other ways, too, and physicists struggle to explain in detail how quarks form mesons and baryons.

Strange quarks

This uncertainty has led to proposals that other forms of matter might exist. As early as the 1980s, Edward Witten, a mathematical physicist at Princeton University, suggested that light quarks could combine with their heavier cousins, such as strange quarks, in unusual ways (see Figure 6.5). These quarks would develop into large amorphous blobs, gathering more and more particles in a small space. Witten called them 'quark nuggets'. Bryan Lynn, a theoretical physicist at University College London, and others later expanded this to more examples such as 'strange baryon matter' and 'chiral liquid drops'.

Such exotic clumps of familiar elementary particles would be as dense as a neutron star – a teaspoon of which weighs as much as a

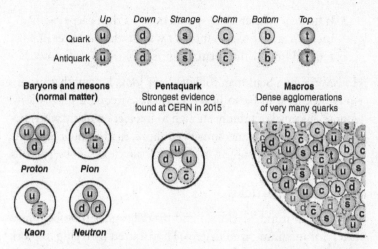

FIGURE 6.5 How quarks might form 'macros': in ordinary matter, quarks are bundled in twos and threes. If more could clump together, they could form ultra-massive particles that could account for the universe's elusive dark matter

mountain. Some researchers have called them 'macros' because we would measure their masses in kilograms and tonnes rather than the vanishingly small units usually employed for particles. Macros would not be capable of sustaining fusion and therefore could not shine; and they would be too small to reflect or absorb much light from elsewhere. In short, they would be almost invisible.

It sounds like the perfect recipe for dark matter, but physicists had previously discounted the idea, for two reasons:

1 If macros are compact objects about as heavy as our Sun, similar to brown dwarfs or black holes, then they would have to outnumber visible stars in order to account for dark matter. If so, macros would show up by bending the light reaching the Earth from stars, an effect known as gravitational lensing.

2 If nuclear matter were spread out in a thin carpet across the universe, it would interact with itself and other matter, and hinder the formation of galaxies as we know it.

However, when Starkman took a closer look, he saw that macros would not have to be so heavy as to cause frequent gravitational lensing, or spread out thinly enough to interact regularly with anything. Clumped into medium-sized drops, neither huge nor tiny, they would be compatible with existing cosmological observations.

Spaceborne particles

With this in mind, Starkman and his colleagues have begun to search for medium-sized macros. They started by trying to work out where macros at the lighter end of the allowed mass scale might have already appeared, potentially leaving tracks in buried minerals or in the plastic sheets mounted on the Skylab space station to catch spaceborne particles. The absence of signals from these and other experiments allowed Starkman to constrain the range of allowed macro masses to between roughly 50 grams and the mass of Mount Everest.

FIGURE 6.6 The Pierre Auger cosmic ray observatory in Argentina, which is used to record oddball matter raining down on Earth.

Starkman's collaborator David Jacobs, from the University of Cape Town in South Africa, now hopes to hear the impact of macros passing through the ocean using marine hydrophones – normally employed to study whales or track illegal nuclear weapons tests. He also plans to study data from cosmic ray detectors, as a macro hitting the Earth's atmosphere would produce a characteristic light signal.

The best bet might be a little farther from home. The last Apollo mission left a network of four seismometers on the Moon, which could betray macro impacts. These seismometers are fairly crude and planetary geologists plan to replace them with better kit: Bruce Banerdt of NASA's Jet Propulsion Lab in Pasadena, California, and his colleagues have drawn up plans for a more sensitive lunar seismic detection network.

Discovering these tiny effects would have a huge impact. It would mean that the range of exotic particles previously invented and sought by physicists may simply not exist, and show that the ordinary elementary particles we know and love can come together in some extraordinary ways.

What happens next in the hunt for dark matter?

Dark matter is at a crossroads. Few problems have received more attention from physicists and astronomers in recent years than trying to discover what it is and how it works. So far, there are few concrete facts, just educated guesses.

That could soon change. Any of the types of dark matter search – direct detection in deep mines, indirect detection with space telescopes, or the Large Hadron Collider – could be near a breakthrough. Could two experiments, called CoGeNT and DAMA/LIBRA, already be seeing hints of dark matter? Is dark matter producing the gamma rays that the Fermi space telescope has observed coming from the centre of our galaxy? There is as yet no consensus on these questions, but time and more data should provide answers.

If dark matter is in fact made up of weakly interacting massive particles (WIMPs), such as particles similar to those predicted by supersymmetry, success could be just around the corner. On the other hand, if no such signals appear in the coming decade, physicists are going to have to throw out much of what they think they know about dark matter and dream up new possibilities. Perhaps dark matter is entirely inert, and does not interact at all with normal matter. If so, it will never be detectable by any of the experiments physicists have been designing – the dark matter hunter's worst nightmare.

Interview: Shooting for a beam of dark light

Does dark matter have its very own dark forces? The only way to find out is to hunt them down, says Tim Nelson, a physicist at the SLAC National Accelerator Laboratory at Menlo Park, California.

Why do you think there is a fifth force?

The four fundamental forces of physics – gravitational, electromagnetic, strong nuclear and weak nuclear – are pretty well understood. But there's always the chance that there's another one we just haven't noticed, perhaps because it's incredibly weak. People have been looking for a new one for a long time. But the forces we're looking for now are different in that they act primarily on dark matter. Just as regular matter consists of a range of particles and forces, I'm motivated by the idea that dark matter is just the lightest, most stable component of an undiscovered 'dark sector' of particles and forces.

Is there any reason to think such a dark sector exists?

It makes increasing sense to consider it. We know dark matter exists, that it interacts gravitationally – in other words, it has mass – and that the vast majority of it is probably embodied in a particular type of particle. Scientists latched on to the idea that dark matter is mostly composed of particles called weakly interacting massive particles. But searches for those WIMPs – with underground detectors and the Large Hadron Collider, for example – haven't turned up anything, and we're running out of room where we might find them. So if dark matter isn't just WIMPs, one solution is that there are different sorts of dark particles that interact with each other via their own set of forces.

Does this mean that dark matter could be pretty diverse stuff?

Yes. The standard model of particle physics has lots of particles, with a set, including the photon, that carries the forces. This ordinary matter only accounts for about one-sixth of the universe's matter. The rest is dark matter, so why wouldn't it be as diverse? If you open that conceptual door, you're suddenly looking at a lot of new possibilities. But to help us get started, we're just considering the simplest option at the moment, which is a dark force analogous to electromagnetism, so we came up with the term 'dark photons'.

How do you go about hunting for dark photons?

The theory is that dark photons mix with regular photons by a process called kinetic mixing. That means a dark

photon can turn into a regular photon, and vice versa –
though most likely at some very, very low rate. So, in principle, if you have an experiment where you produce lots
of high-energy photons, you'll also produce dark photons
at a much lower rate.

How do you detect dark photons?

Dark photons can't be massless like regular photons. If
they were, it would contradict our understanding of how
dark matter behaves. In fact, they could have an incredibly
wide range of masses. That means that, although we can't
see the dark photons directly, we can hunt for them the
same way we hunt for any other particle that has mass.

Are you trying this already?

Yes, our Heavy Photon Search experiment at the Thomas
Jefferson National Accelerator Facility uses a beam of
high-energy electrons that we fire into a tungsten foil target. When you do that – and the electrons suddenly hit
that obstacle – you get deceleration radiation. That radiation is essentially a beam of photons, and, if dark photons exist, the collisions will radiate those too, at a lower
rate. What happens next depends on whether or not dark
photons are the lightest particle of the dark sector. Our
experiment assumes that they are, which means they must
decay via kinetic mixing to regular matter such as electron–positron pairs, which we can detect.

What if dark photons are heavier than you think?

We presume that the lightest dark matter particle makes up
the bulk of dark matter. If the dark photon is not the lightest
particle in the dark sector, then, instead of decaying back to

regular matter, it will pretty much always decay to dark matter. That means we can't see it with our current experiment, but it actually leads to some interesting possibilities. If I have an experiment with a really thick tungsten target and create a bunch of dark photons that are all moving really fast, they are all going to decay to dark matter – so I have now essentially created a dark matter beam. We will have lost the ability to detect dark photons, but gained the ability to detect dark matter itself. It's a win–win situation.

Tell us more about the dark matter beam.

The cool thing about this is that it would produce dark matter at high energy. The direct-detection experiments we have for dark matter, such as LUX and CDMS, are trying to detect dark matter orbiting our galaxy at relatively low velocities. When it bumps into the detector, it deposits only a very small amount of energy that is really hard for us to spot. That's why we have to bury these detectors in mines – the surrounding earth screens out many interfering signals. But if I had a high-energy dark-matter beam, I could just point it at a standard particle detector.

What would definitive detections of dark matter mean for humanity?

It would be like the Copernican Revolution – another confirmation that we are not at the centre of the universe, and that what we thought was the entire universe is just a tiny slice. It's one thing to be intellectually convinced of that, as we are, but another to confront it face to face.

7
Dark energy

Our cosmos is dominated by a mysterious force called dark energy, which was first suspected as a result of Einstein's equations seven decades before it was actually detected. Dark energy, along with dark matter, completes the Einsteinian universe.

Dark energy: still the greatest cosmic mystery

It comprises two-thirds of the cosmos but it just keeps us guessing. Is dark energy a new field, a new force or the power of our own ignorance?

A couple of decades ago, we noticed that some mysterious agent is pushing the universe apart. This agent is everywhere but we cannot see it. Despite the fact that it makes up more than two-thirds of the universe, we have no idea what it is, where it comes from or what it is made of. We do at least have a name for this most enigmatic of beasts: dark energy. And in recent years the hunt for it has been gathering pace, with new sky surveys to look for signs of it among exploding stars and ancient galaxy clusters. Space missions and gigantic Earth-based telescopes will soon join the search, while some physicists are trying to capture dark energy in the lab.

So far, we know three things about dark energy:

1 It pushes. In 1998 the unexpected dimness of certain supernova explosions told us they were farther away than we expected. Space seems at some point to have begun expanding faster, as if driven outwards by a repulsive force acting against the attractive gravity of matter.

2 There is a lot of it. The motion and clustering of galaxies tells us how much matter is abroad in the universe, while the cosmic microwave background radiation allows us to work out the total density of matter plus energy. That second number is much bigger. About 68 per cent of the universe is in some non-material, energetic, pushy form.

3 It makes excellent fuel for the creative minds of physicists, which have turned it into hundreds of different and fantastical forms.

Why does dark energy push?

We are used to gravity pulling things together, so it is disconcerting to find a cosmic force that instead pushes outwards. Vacuum energy is strange in other ways, too: as space expands, there is more and more of the stuff, accounting for more and more energy. It behaves like a spring or rubber band that gains energy when you stretch it out. In other words, vacuum energy is under tension.

But shouldn't something under tension exert a pull rather than a push? At this point our intuition fails. In general relativity, what tells you how gravity behaves is not just energy, but also pressure. High pressure generates attractive gravity and, conversely, tension means repulsion.

Clearly, this is not a satisfying explanation. High-school physics takes you halfway there, to the idea that dark energy has tension; and then a miracle of relativity occurs, making that tension repulsive. Somewhere on the great quest to find out what dark energy really is, perhaps we will stumble on a good way to explain why it pushes.

The cosmological constant

The tamest of these hypothetical beasts is the cosmological constant, and even that is a wild thing. It is an energy density inherent to space, which within Einstein's general theory of relativity creates a repulsive gravity. As space expands, there is more and more of it, making its repulsion stronger relative to the gravity of increasingly scattered matter. Particle physics may provide an origin for it, in virtual particles that appear and disappear in the bubbling, uncertain quantum vacuum. The problem is that these particles seem to have far too much energy – in the simplest calculation, about 10^{120} joules per cubic kilometre.

That catastrophic discrepancy leaves room for several alternative theories. Dark energy could be some kind of energy field that permeates space, changing over time and perhaps even clumping in different places. It might be a modified form of gravity that repels at long range. In one theory, dark energy takes the form of radio waves trillions of times larger than the observable universe. Or it could be something even more exotic than that.

Astronomers want to find out whether dark energy is changing over time. If it is, that would rule out the cosmological constant, whose density would remain unchanged. A new kind of energy field, by contrast, may become slowly diluted as space stretches, or it may intensify, pumped up by the universe's expansion. In most modified theories of gravity, dark energy's density is also variable. It might even go up for a while and then down, or vice versa.

The fate of the universe hangs in this balance. If dark energy remains steady, most of the cosmos will accelerate off into the distance, leaving us in a small island universe, forever cut off from the rest of the cosmos. If it intensifies, it might eventually shred all matter in a 'big rip' or even make the fabric of space unstable here and now. Our best estimate today, based mainly on supernova observations, is that dark energy's density is fairly stable.

Chasing shadows

The Dark Energy Survey aims to look for several tell-tale signs of dark energy over a wide swath of the sky. It uses the 4-metre-wide Víctor M. Blanco telescope at the Cerro Tololo Inter-American Observatory in Chile, attached to a specially designed infrared-sensitive camera.

The survey will catch many more supernovae. The apparent brightness of each stellar explosion tells us how long ago it happened. During the time the light has taken to reach us, its wavelength has been stretched, or redshifted, by the expansion

SUPERNOVAE

Distant type Ia supernovae are dimmer than expected, suggesting they are farther away.

COSMIC MICROWAVE BACKGROUND

With only matter's gravity at work, the universe should be curved. Patterns in the Big Bang's afterglow suggest it is nearly flat.

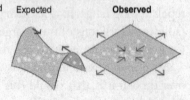

GRAVITATIONAL LENSING

Images of distant galaxies are less distorted by intervening matter than we expected. A repulsive force seems to be stopping matter clumping.

ACOUSTIC IMPRINTS

Sound waves rippling through the early universe gave galaxy superclusters a typical scale. Far-off superclusters appear smaller than expected – and so are farther away.

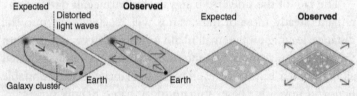

FIGURE 7.1 The mystery of dark energy: many lines of evidence indicate that something is countering gravity's pull and accelerating the universe's expansion

of space. Put these two things together and we can plot expansion over time.

The survey will also draw an intricate sky map that marks the positions of a few hundred million galaxies and their distances from us. Sound waves reverberating around the infant cosmos gave vast superclusters of galaxies a characteristic scale. By measuring the apparent size of superclusters, we can gain a new perspective on the expansion history of the universe (see Figure 7.1).

The map will also reveal dark influences on smaller scales. Dark energy hinders galaxies from coming together to form clusters. The survey team will count clusters directly, and also follow their growth using gravitational lensing, an effect that happens when clusters bend light passing through them from even more distant cosmic objects (see also Chapter 2). Cluster growth could reveal modified forms of gravity, as it probes intermediate distance scales where gravity might start to change its nature. All these various measurements should give us an idea of how dark energy has changed over time.

Mega-scopes

The DES is just the leader of a group of dark energy hunters. The Large Synoptic Survey Telescope, a US-led project, is due to open its great eye in around 2021. Other mega-scopes such as the European Extremely Large Telescope in Chile should go into action at around the same time, along with the Square Kilometre Array: a huge cosmic radio receiver based in Australia and South Africa, which will trace cosmic structure through the radio glow of hydrogen clouds. In 2020 the European Space Agency plans to launch a dark-energy hunting space mission called Euclid that will trace gravitational lensing and galaxy clumping to even earlier cosmic times. NASA's Wide-Field Infrared Survey Telescope should follow a few years later.

This chase through space will be thrilling, but the quarry may still elude us. If we find that dark energy maintains a near-constant density over time, that would seem to support the cosmological constant, but it would not rule out some quintessence fields that just happen to have a near-constant density. Consequently, some physicists are laying traps for dark energy here on Earth. If you introduce a new field or particle to be dark energy, then it will also act as the carrier of a new force – separate from gravity, electromagnetism and nuclear forces. But we don't see such a thing affecting planetary orbits in our solar system. Many theorists get out of this by adding a screening mechanism that weakens the fifth force in comparatively dense environments such as the solar neighbourhood.

Collective quantum waves

Clare Burrage at the University of Nottingham, UK, realized that we could search for this effect in the lab. She and her collaborators aim to use a cloud of cold atoms called a Bose-Einstein condensate, which oscillate together in a collective quantum wave. Some forms of dark energy should just slightly slow down the frequency of this oscillation, so the team plans to split a condensate in two and place a dense object near one of the halves. If the object screens out dark energy, waves in the two halves will get out of sync, and when brought back together the two condensates will interfere. A version of this experiment has already been carried out by Paul Hamilton's team at the University of California, Berkeley – but with, so far, no sign of screened dark energy.

At the University of Washington in Seattle, the Eöt-Wash torsion pendulum experiment is probing other forms of cosmic repulsion. In one theory, extra dimensions of space less than a millimetre across can play host to dark energy. That would also increase the strength of gravity on these scales. A type of screened quintessence called the symmetron would generate a similarly small-scale extra force – a tiny effect that the subtle twistings of the Eöt-Wash pendulums should expose.

Meanwhile, in 2016 Michael Romalis at Princeton University and Robert Caldwell at Dartmouth College in Hanover, New Hampshire, proposed that if ordinary photons or electrons can feel dark energy even very faintly, then a magnetic field on Earth should generate a tiny electrostatic charge. This effect is potentially simple to spot, although any apparatus designed to do it would need to be very precise.

Few imagine that the hunt will be over soon. After two decades of puzzlement, we have still have no clue to dark energy's identity. But on the bright side, we do have some clues to where the clues may lie.

Is dark energy an illusion?

Cosmology's standard model brings order to everything, from the patterns in the Big Bang's afterglow to the evolution of galaxies. But it can only do so using the powers of dark matter, the additional unseen stuff amounting to a quarter of everything in existence, and dark energy, the mysterious, expansionist force that seems to dominate the universe. But dark energy's power, some claim, is a mere illusion.

At the heart of this conflict lies the cosmological principle, which states that the universe is more or less the same no matter where you are or in whichever direction you look. It proved a boon when trying to extract workable models of the universe from the notoriously intractable equations of general relativity. On one side of these equations are mathematical terms for things that warp space and time: matter, energy and pressure. On the other side are descriptions of their effects: how fast space-time is expanding and its curvature. (Space-time can be folded in on itself as if on the surface of a four-dimensional sphere, called positive curvature, or it can be folded outwards with negative curvature, or it can be 'broadly flat' with zero curvature.)

Building a cosmological model means balancing all these terms for the entire universe – the right amount of stuff to produce the right expansion and curvature. To make that job more manageable, cosmologists assume a uniform universe, where matter and energy are evenly spread, and an overall average curvature that does not change in time or space. This leads to solutions to the equations that describe a smooth, expanding cosmos: the standard model of the Big Bang universe.

The model has had to be tweaked a couple of times. The discovery that galaxies and clusters of galaxies were rotating too fast for the amount of visible matter they contained was solved by adding dark matter to the mix. Then, in 1998, we found that expansion is accelerating. Add a constant or near-constant term to the model on the matter and energy side and you can make the equations balance out, reproducing a flat universe with accelerating expansion. But no one knows what that term really is, even though the model requires it to account for almost 70 per cent of the total energy in the universe.

This orthodoxy is what the rebels are seeking to undermine. Abandon the assumption that the universe is uniform and unchangingly flat, they say, and you can eliminate dark energy and possibly dark matter too.

Seeds of galaxies

Already in the cosmic microwave background, you see the seeds of the galaxies and clusters of galaxies that gravity's pull has constructed over time. As the universe has evolved, a web of over-dense regions has gradually formed, with huge under-dense voids opening up between them.

What effect has this changing distribution of matter had on the space-time around it? The extra mass of galaxies and galaxy clusters should increase the curvature of nearby space-time, making it more positive (see Figure 7.2). Meanwhile, voids will

A change in the universe's geometry could create the illusion that its **expansion is accelerating**, an effect usually ascribed to **dark energy**.

There are three basic possibilities for how matter makes space-time curve:

High matter density ————————————————→ Low matter density

Positive curvature
(Closed geometry)
Parallel light beams converge.

Zero curvature
(Flat geometry)
Parallel light beams
stay parallel.

Negative curvature
(Open geometry)
Parallel light beams
diverge.

The clumping of matter into structures such as galaxies over time could have changed the universe's overall geometry:

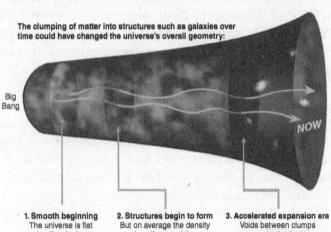

Big
Bang

NOW

1. Smooth beginning
The universe is flat
and uniformly dense.

2. Structures begin to form
But on average the density
is uniform and the
geometry remains flat.

3. Accelerated expansion era
Voids between clumps
come to dominate,
forcing the overall
curvature below zero.

Light beams
stay parallel.

Light beams stay
parallel.

Light beams diverge in the voids,
making the universe look
bigger without dark energy.

FIGURE 7.2 Bending space

cause their local space-time to warp the other way, giving it a negative curvature.

According to Thomas Buchert at the University Claude Bernard Lyon 1, France, the resulting curvature might make it look as if there is additional stuff there, giving the illusion of dark matter. But this is unlikely, as Buchert emphasizes. Other lines of evidence lead to the conclusion that dark matter must exist, and some of them (such as patterns of sound waves imprinted on the cosmic microwave background in the early years of the universe) cannot be explained away in this way.

While dark matter may be difficult to banish, the rebels have set their sights on the bigger target of dark energy. Syksy Räsänen of the University of Helsinki, Finland, suggests that these local 'backreaction' effects between matter and space-time could also change the geometry of the universe as a whole. As matter clumps into ever denser, more compact structures, the proportion of the universe that is void increases, pushing its overall average curvature into negative territory. In a universe whose curvature goes negative over time, light rays will become distorted and things will look more distant than in a flat space-time. So you can build models in which there is no accelerated expansion and no dark energy.

Blown apart

This does away with another problem, too. In the standard model, it is difficult to explain why dark energy's effects happen only about 5 billion years ago, some 9 billion years into the universe's history. This is a crucial point: had dark energy dominated earlier, the universe would have blown apart so quickly that we would have had no galaxies, no life and no physicists wondering about appropriate models of the universe. With backreaction, there is nothing to explain: 5 billion years ago is the point in the progressive evolution of structures when voids begin to dominate and the overall curvature goes negative.

Räsänen is moderate in his position, concluding that the case for dark energy is not established beyond reasonable doubt. Others, such as David Wiltshire of the University of Canterbury in Christchurch, New Zealand, are more forthright, claiming that there is no dark energy. Wiltshire, Buchert and fellow backreactionists say they can fit existing observations to models that do not include dark energy.

The best way to find out for sure would be to build general relativistic models that simulate the evolution of a realistic lumpy universe. Until recently, the huge computational demands of such an endeavour made it impossible. But now two teams, led by Glenn Starkman of Case Western Reserve University in Cleveland, Ohio, and Tom Giblin at Kenyon College in Gambier, Ohio, have begun to make numerical simulations of the universe using the full power of general relativity. The preliminary results suggest that backreaction effects do affect local expansion rates, but that they are not strong enough to alter the overall curvature of the universe by much and generate the observed acceleration.

Wiltshire points out that these models do not yet allow the average curvature of space-time to evolve over time. And Starkman himself cautions that the models are still crude: the distribution of matter is still not fine-grained enough to be entirely realistic, and matter is modelled as a fluid, not particles. For now, at least, the backreactionists are not giving up the fight against the dark energy model.

Discovering a cosmic mystery

Along with Brian Schmidt and Saul Perlmutter, Adam Riess won the 2011 Nobel Prize in physics for the discovery that the expansion of the universe is speeding up. Educated at the Massachusetts Institute of Technology and Harvard University, Riess now works at Johns Hopkins University and the Space Telescope Science Institute, both in Baltimore, Maryland.

What was the Nobel prize-winning discovery that you, Schmidt and Perlmutter made?

We were two teams of astronomers who observed nearby and distant supernovae and used them to infer the amount of expansion of the universe at different times in its history. We determined that the universe, contrary to expectations, was not slowing down in its expansion – it was actually speeding up.

Dark energy is supposed to be the inherent energy of space-time, something we still don't fully understand. Wouldn't it have been odd to receive a prize for having discovered a mystery?

Absolutely. The acceleration of the universe is the smoking gun of something. It could be that gravity on larger scales doesn't operate the way we think it does. The cleanest thing we can say is that the universe's expansion is accelerating and that's a great surprise.

Einstein had the idea that space-time has an intrinsic energy density that does not change with time, called the cosmological constant – but he later called this

notion his 'biggest blunder'. Is your work a vindication for him?

It's an impressive vindication of Einstein's general relativity. All these decades later, when we see very exotic phenomena in the universe, they can be fully accommodated, even expected, in his theory.

It wasn't that far back that astronomy and astronomical observations were not even considered for the Nobel Prize.

Right. I could certainly name discoveries from the past from cosmology that are completely Nobel-worthy: detecting the expansion of the universe or the scale of the universe, and the observations that indicate the presence of dark matter or some kind of additional gravity. These are fundamental to our understanding of physics.

8

Beyond relativity

To really understand black holes, the beginning of time and the true nature of space, we need a theory that somehow blends general relativity and quantum mechanics — two ideas that seem to be fundamentally incompatible.

The odd couple

Some day, the two great pillars of modern physics must be joined together. General relativity and quantum mechanics have had great success separately, but they seem incompatible. The standard model of particle physics, in a quantum framework, describes most of the fundamental forces of nature in terms of flitting particles, while relativity describes gravity in completely different terms using curved space-time.

The pillars clash in other ways: wherever we find a physical situation where the two theories are both important – such as the event horizon of a black hole – they fail to work together (see Chapter 3). A quantum theory of gravity seems to be necessary to deal with the first moments of the Big Bang, and probably to understand the nature of space and time. But quantum gravity has stumped us all. Einstein himself became extremely unproductive in his later years as he sought such a 'theory of everything'.

To understand the problem, we must start with a fundamental tenet of quantum physics. Heisenberg's uncertainty principle embodies the 'fuzziness' of the quantum world. It allows particles to borrow energy from empty space, to come into existence as short-lived 'virtual' particles. They must pay this energy back by disappearing again – and the more they borrow, the more quickly that must happen.

One can imagine an electron, a photon or any other particle going to town, taking out many zero-interest loans in succession. As a result, calculating even a simple quantum process – an electron travelling from left to right, say – becomes enormously complex. In the words of physicist Richard Feynman (1918–88), we must 'sum over all possible histories', taking into account the infinite variety of ways virtual particles can be produced (see Figure 8.1).

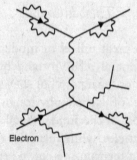

Electron

Particles such as electrons can interact by producing
and exchanging massless photons in countless ways,
often resulting in infinities in the calculations.

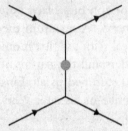

The situation is saved by the existence of heavier
particles – the **W, Z and Higgs bosons** – that aren't
so easily produced, cancelling out the infinities.

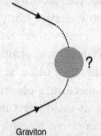

Graviton

Performing the same trick with graviton interactions
requires a particle so massive it acts like a
black hole – and all calculations are off again.

FIGURE 8.1 The infinite problem: gravitons are conjectured quantum
particles of gravity, but theories incorporating them tend to be unruly

Quantum electrodynamics

Sometimes when you do this sum, you get a finite answer – the theory makes a prediction that can be tested. Quantum electro-dynamics does this for situations like the moving electron. At other times, the sum blows up and you get infinity. The history of applying quantum theory to nature's forces is a history of getting to grips with these unruly infinities.

One case is beta decay – when a neutron spontaneously spits out an electron and a neutrino, leaving a proton behind. The quantum theory of beta decay gave infinities – until physicists developed electroweak theory, which combines the electromagnetic and weak nuclear forces. Electroweak theory tamed the mathemat-ics by adding undiscovered massive particles, the W, Z and Higgs bosons, which cancelled out the infinities. Fortune favoured this brave conjecture: the W and Z bosons were discovered at CERN in 1983, with the Higgs following in 2012. Its success led many physicists to believe this strategy was something like a general pre-scription for developing quantum theories: if your model pro-duces infinities, just add extra particles to solve the problem.

Suppose, then, that gravity is made of quantum particles called gravitons, as light is made of photons. Following the uncertainty principle, gravitons borrow energy to make other, virtual gravi-tons, and as we sum over all possible histories, the calculations rapidly spiral as expected, into a chaos of infinities. But if we try to fix the infinities by adding more particles, it doesn't work. It requires the invention of a new particle with a mass 10 billion billion times that of a proton. As ever, the larger the amount of energy borrowed, the more quickly it must be paid back, so these fixer particles are very short-lived. This means they can-not get very far, and so occupy only a minute amount of space.

But general relativity says that mass bends space-time. Con-centrate enough mass into a small area and a black hole will form. And this is exactly the guise our new particle takes, a microscopic black hole containing a singularity of infinite density and infinite curvature in space-time. Nature plays a cruel joke on us: our scheme to eliminate one sort of infinity creates another.

Attempts to get round this fundamental roadblock have led us to destinations such as string theory, which assumes that all particles are manifestations of more fundamental vibrating strings. When we start summing over all possible histories of these 'fluffier' objects, the hard infinities produced by virtual particles drop away almost by magic (see 'A theory of everystring' below). Another idea is loop quantum gravity, which suggests that space-time itself is chopped up into discrete blocks. This pixellation imposes an upper limit on the amount of energy any particle can borrow, again rendering calculations finite.

These two candidate unified theories are in many ways the most conservative extensions of current models: both attempt to preserve as much of the theoretical underpinnings of quantum mechanics and general relativity as possible. What about more esoteric ideas, such as changing the rules of the existing game? For instance, if we were to treat space and time separately again, rather than lumping them into one combined space-time, that might provide some wiggle room (see the box on 'Horava gravity' later in this chapter). Could we make progress by abandoning the cornerstone of general relativity – the principle of equivalence (see 'Differently equal' later in this chapter)? Or do we need a more fundamental shift of perspective, seeking the nature of reality in pure numbers?

Relativity and quantum mechanics both tally so well with reality in their respective spheres that it is extremely difficult to formulate something better, but few physicists would care to consider an even more radical possibility: that quantum mechanics and general relativity cannot be unified, and reality has no single, consistent logical underpinning.

A theory of everystring

An all-embracing theory of physics that unifies quantum mechanics and general relativity might be able to describe everything in the universe, from the Big Bang to subatomic particles. Do we finally now have a leading candidate for this theory of everything?

At the end of the nineteenth century, atoms were believed to be the smallest building blocks of matter. Then it was discovered that they have a structure: a nucleus made of protons and neutrons, with electrons whizzing around it (see Figure 8.2). In the 1960s the atom was divided even further when it was theorized, then confirmed by experiments, that protons and neutrons are composed of yet smaller objects, known as quarks. Do these layers of structure imply an infinite regression? All the theoretical and experimental evidence gathered so far suggests not: quarks really are the bottom line. We now believe that quarks are the fundamental building blocks of matter, along with a family of particles called the leptons, which includes the electron (see Figure 8.3).

The quarks and leptons that make up matter seem very different from bosons, the particles that carry nature's forces. So it came as a great surprise in the 1970s when theorists showed that it is possible to construct equations that stay the same when you swap the two around. This suggests a new symmetry of nature. Just as a snowflake's underlying symmetry explains why it can look the same even after you rotate it, so the unchanging equations are down to a new symmetry, called supersymmetry. One consequence is that every particle in the standard model has a supersymmetric partner. No one has found one yet.

FIGURE 8.2 The classical Bohr model of an atom. In quantum physics the electrons do not follow defined orbits, but instead trace out clouds of probability around the nucleus.

Supersymmetry and supergravity

Theorists, however, remain enamoured with supersymmetry (see also Chapter 6) because it predicts gravity. According to the mathematics of supersymmetry, the act of turning a particle

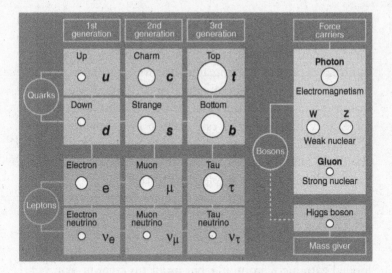

FIGURE 8.3 The standard model: this is our best understanding of the building blocks of matter and the forces that glue them together

into its supersymmetric partner and back again is identical to moving it through space-time. This means that supersymmetry offers a connection between the properties of quantum particles and space-time, making it possible to incorporate gravity, too. The resulting theory is known as supergravity. The mathematics of supergravity have an unexpected consequence: space-time can have no more than 11 dimensions.

The idea of adding extra dimensions to the universe goes back to an early attempt at unifying the forces of nature. In the 1920s the German mathematician and physicist Theodor Kaluza (1885–1954) rewrote Einstein's theory as if there were five space-time dimensions, which gives the gravitational field extra components that look like the electromagnetic field. Why don't we see the fifth dimension? In 1926 the Swedish physicist Oskar Klein (1899–1974) supposed that the fifth dimension is not like the other four, but is instead curled up into a circle too small to see. Imagine an ant on a narrow tightrope: as well as walking along the rope, the ant can choose to walk around its circumference at any point. But viewed from a distance much, much larger than the ant's size, the rope looks very different: it is essentially a one-dimensional line. Klein's calculations showed that the extra dimension should be about 10^{-35} metres across, far too small to probe even with the most powerful particle accelerators today.

Kaluza and Klein's idea lay dormant for many years, until supergravity prompted a revival, with now up to seven curled-up dimensions. Could these extra dimensions describe the strong, weak and electromagnetic forces? At first, supergravity looked extremely promising, but problems crept in. For a start, 11-dimensional supergravity has trouble describing how quarks and electrons interact with the weak nuclear force. Even more serious is a problem that has dogged all other attempts to reconcile gravity and quantum field theory: when you use supergravity's equations to calculate some processes, the answer is infinity.

String vibrations

Attention turned to a rival approach called superstring theory, where the building blocks of matter are not point-like particles but one-dimensional strings that live in a universe with ten dimensions. Just like violin strings, they can vibrate in various modes, each one being a different elementary particle. Certain string vibrations can describe gravitons, the hypothetical carriers of the gravitational force.

This looked like a theorist's dream. It could deal happily with the weak force, unlike 11-dimensional supergravity. Also, superstring theory looked just like general relativity when the graviton energy was set sufficiently small. But the most important feature was that the infinities and anomalies that had plagued previous attempts to apply quantum field theory to general relativity no longer existed. Here was a consistent way to unify gravity with quantum mechanics.

However, after the initial euphoria, doubts began to creep in. For a start there are not one, but five, mathematically consistent superstring theories, each competing for the 'theory of everything' title. Supersymmetry says that the universe has a maximum of 11 dimensions, yet the mathematics of superstring theory states that there should be ten. And why stop at one-dimensional strings? Why not two-dimensional membranes, which might take the form of a sheet or the surface of a bubble?

Are superstrings the same as cosmic strings?

The answer is no – or possibly yes.

Cosmic strings are hypothetical defects in space-time, which would appear as lines of concentrated energy billions of light years long, so dense that a piece 1 metre long weighs as much as a continent. This is in contrast to superstrings, which are 10^{-35} metres long, and weigh as much as whatever subatomic particle they are being at the moment.

We haven't seen evidence for cosmic strings, but some cosmologists think they could have formed when the very early universe cooled, and the quantum vacuum went through a series of phase transitions, akin to melting ice. No superstrings were needed there. But cosmic strings might also have formed in another way, if the rapid expansion of space in the era of inflation was able to grab primordial superstrings and stretch them to cosmic proportions.

Supermembranes

It turns out that supersymmetry and membranes do go together. It was calculated in 1987 that 'supermembranes' can exist in an 11-dimensional space-time dictated by supergravity. For many years there were two camps: string theorists with their ten-dimensional theory, and the membrane theorists working in 11 dimensions.

All this was brought together in 1995 under one umbrella called M-theory, by Edward Witten at the Institute for Advanced Study in Princeton. 'M', he says, stands for magic, mystery or membrane according to taste. Witten showed that the five different string theories and supergravity were different facets of M-theory (see also Chapter 5).

M-theory and its membranes were able to do things strings could not. In 1974, Stephen Hawking showed that black holes can radiate energy due to quantum effects, meaning that they have temperature and another thermodynamic property called entropy, which is a measure of how disorganized a system is. Hawking showed that a black hole's entropy depends on its area. It should be possible to work out its entropy by accounting for all the quantum states of the particles making up a black hole, but all attempts to describe a black hole in this way had failed – until M-theory came along, exactly reproducing Hawking's entropy formula.

In 1998 the Argentinian physicist Juan Maldacena showed that everything happening inside a universe in M-theory can be described by particles interacting on the boundary of that universe. This 'holographic principle' could mean that we are just shadows on the boundary of a higher-dimensional universe.

The precise way that extra dimensions are curled up is what dictates the appearance of our four-dimensional world – including how many generations of quarks and leptons there are, which forces exist, and the masses of the elementary particles. A puzzling feature of M-theory is that there are many (perhaps infinitely many) ways of curling up these dimensions, leading to a plethora of possible universes. Some may look like ours, with three generations of quarks and leptons and four forces; many will not. But from a theoretical point of view they all seem plausible. So there could be multiple universes out there with different laws of physics, and one of these universes just happens to be the one we are living in.

Is M-theory the final theory of everything? It is hard to test this idea. Some generic features such as supersymmetry or extra dimensions might show up at collider experiments or in astrophysical observations, but the variety of possibilities offered by the multiverse makes precise predictions difficult. Are all the laws of nature we observe derivable from fundamental theory or are some mere accidents? These key issues may remain unresolved for some time. Finding a theory of everything is perhaps the most ambitious scientific undertaking in history. No one said it would be easy.

The string theory revolutionaries

In 1990 the string theory pioneer Edward Witten won the Fields Medal, the mathematics equivalent of the Nobel Prize. This shows just how closely mathematics and string theory tie together.

Juan Maldacena is another of today's most influential physicists. His work showed that the physics inside a region of space can be described by what happens on its boundary. While his idea originated in M-theory, it has gone on to revolutionize many areas of theoretical physics.

A radical new route to a theory of everything

For decades, the hunt for a theory to unite relativity and quantum theory has been based on string theory. But there are alternatives.

The most advanced non-string theory approach to a theory of everything is known as causal dynamical triangulations (CDT). Developed by by Renate Loll of Utrecht University in the Netherlands and her colleagues Jan Ambjorn and Jerzy Jurkiewicz, CDT models space-time as made up of tiny, identical building blocks (see Figure 8.3). These are higher-dimensional analogues of triangles, called 4-simplices. Governed by quantum mechanics, the triangles perpetually rearrange themselves into new configurations, each of which has its own curvature.

Just as you can glue six equilateral triangles together at a point to make a piece of flat space, so CDT can produce flat, positively curved or negatively curved space-time by allowing different numbers of 4-simplices to meet at a point. The triangles are not physical objects but a mathematical and computational tool – and they lead to compelling results.

The crucial step in deriving space-time on a large scale is to sum over all possible configurations of these triangles. This is in keeping with the spirit of Richard Feynman's approach to quantum mechanics, in which every possible path of a particle must be added up to calculate how it gets from A to B. Back in the late 1970s, a similar approach to space-time was taken by Stephen Hawking at the University of Cambridge, but it produced universes with either no dimensions or an infinite number of them.

Loll's insight was to insist that a fixed ordering of cause and effect be incorporated in the way the triangles can arrange themselves. Now when she performed the calculation, what emerged were three dimensions of space and one of time,

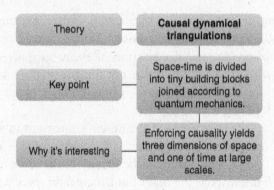

FIGURE 8.4 In a nutshell: causal dynamical triangulations (CDT)

making a smooth, expanding universe like the one in which we live, governed by the laws of general relativity and consistent with standard cosmology.

The result suggests that causality may explain why we live in a four-dimensional universe. But in CDT, space-time is four-dimensional only at large scales. At the tiniest scales, the model comes up with only two dimensions, and it produces a space-time with a fractal structure – rough and jagged, with increasingly detailed structure as you zoom in. Reality may be a bit frayed at the edges.

Horava gravity

Physicists struggling to reconcile gravity with quantum mechanics have hailed a theory – inspired by pencil lead – that could make it all very simple. Space-time is a concept that has served us well, but if physicist Petr Horava is right, it may be no more than a mirage. Horava, at the University of California, Berkeley, wants to rip this fabric apart and set time and space free from each other.

One of the central ideas of relativity is Lorentz symmetry: all observers moving at constant velocity agree on the laws of physics and the position of events in space-time. But what if this symmetry is not fundamental to nature, but something that emerged as the universe cooled from the Big Bang fireball?

In 2009 Horava amended Einstein's equations in a way that removed Lorentz symmetry. To his delight, this led to a set of equations that describe gravity in the same quantum framework as the other fundamental forces of nature: gravity emerges as the attractive force due to quantum particles called gravitons, in much the same way that the electromagnetic force is carried by photons.

Horava gravity can be studied using the same mathematical tools that have been developed for the three other fundamental forces of nature. This is partly why many physicists have avidly taken up the theory. It might help with the long-standing puzzle of dark matter. When Shinji Mukohyama at the University of Tokyo in Japan extracted the equations of motion from Horava's theory, he found that they came with an extra term that is not present in equations derived from general relativity – and that this extra term mimics the effects of dark matter. Depending on its value, you can do away with some dark matter, or even most of it.

Dark energy is a more daunting problem still. The theories of particle physics predict the strength of dark energy to be about 120 orders of magnitude larger than what is observed – but Horava's theory contains a parameter that can be fine-tuned to reduce the vacuum energy predicted by particle physics.

Differently equal

Our hopes of finding a theory of everything depend on upsetting a balance that Einstein cherished. Modern physics is precariously balanced on an enormous coincidence, essential to how we view and define mass. Einstein declared this coincidence to be a principle of nature, and used this 'equivalence principle' as the foundation of his general theory of relativity. But if we want to find some bigger, better theory that can unify gravity with the other forces, the equivalence principle may have to go.

There are several versions of the equivalence principle, but all boil down to one idea: that the effects of gravitational fields are indistinguishable from the effects of accelerated motion. A person standing inside an elevator on Earth is held down by the pull of gravity. Take the lift to deep space and keep it accelerating with a rocket; now the passenger is held down in exactly the same way. In this case, it is the person's inertia that is preventing them from floating upwards. Inertia is the natural resistance of any body to acceleration – the same effect that pushes you back into your car seat when the driver puts their foot down.

The two elevator situations have a common property: mass. But the two masses come from very different places. One, gravitational mass, is something that responds to the pull of gravity, tending to accelerate a body in a gravitational field. The other, inertial mass, is the property of a body that opposes any acceleration. Another way of stating the equivalence principle is to say that these two masses are always numerically exactly the same. If they weren't, objects of different masses could fall to Earth at different rates. In fact, the equality dictates all gravitational motion throughout the universe. If gravitational mass responded just a little more to gravity than inertial mass does to acceleration, then planets would orbit their stars and stars orbit their galaxies just a little faster than they do.

In Einstein's general theory of relativity, what looks like gravity is just uniform motion through warped space. There is no gravity; so gravitational mass is fictitious, and the coincidence behind equivalence disappears. But if gravity is to be brought into the quantum fold along with the other forces, it will need something to latch on to – just as electromagnetism latches on to electric charge. It needs a gravitational mass that is separate and distinct from inertial mass.

Interview: A theory of everything won't provide all the answers

We should not be obsessed with finding a theory of everything, says Lisa Randall, Professor of Physics at Harvard University.

Doesn't every physicist dream of one neat theory of everything?

There are lots of physicists! I don't think about a theory of everything when I do my research. And even if we knew the ultimate underlying theory, how are you going to explain the fact that we are sitting here? Solving string theory won't tell us how humanity was born.

So is a theory of everything a myth?

It's not a fallacy. It's one objective that will inspire progress. I just think the idea that we will ever get there is a little bit challenging.

Isn't beautiful mathematics supposed to lead us to the truth?

You have to be careful when you use beauty as a guide. There are many theories people didn't think were beautiful at the time, but did find beautiful later – and vice versa. I think simplicity is a good guide: the more economical a theory, the better.

Is it a problem, then, that our best theories of particle physics and cosmology are so messy?

We're trying to describe the universe from 10^{27} metres down to 10^{-35} metres, so it's not surprising there are lots of ingredients. The idea that the stuff we're made of should be everything seems quite preposterous. Dark matter and dark energy – these are not crazy ingredients we're adding.

Deeper and more precise

One way to refute Einstein's equivalence principle is to try to prove that the two masses are not actually equivalent – just very, very close. Even the slightest sliver of a difference would mean that general relativity is built on an approximation and that a deeper, more precise theory must exist.

So far, tests of rubidium and potassium atoms in free fall at the University of Bremen's 'drop tower' have shown no deviation from the equivalence principle. The atoms have been found to fall at the same rate to accuracies of 11 decimal places. At the University of Washington in Seattle, meanwhile, Eric Adelberger and his 'Eöt-Wash' team use a high-tech set of scales known as a torsion balance to compare the motions of standard masses made of different elements, including copper, beryllium, aluminium and silicon. They hold the record for test accuracy, with no violations of the equivalence principle to 13 decimal places. And now the French-led MICROSCOPE mission, launched in April 2016, is testing the motions of masses of platinum and iridium in the microgravity conditions of space.

Meanwhile, theorists are picking at a different thread. They point out that no one has yet come up with a convincing explanation of inertia. One thing is for sure: not all of it comes from the Higgs field. While the Higgs field is thought to give fundamental particles such as electrons and quarks their mass, when quarks combine into the heavier particles, protons and neutrons that make up the bulk of normal matter, the resulting mass is roughly a thousand times the summed mass of the constituent quarks. This extra mass comes not from the Higgs mechanism but from the energy needed to keep the quarks together. Somehow, these two effects must combine and latch on to something else to create the property of a body's resistance to acceleration.

The accelerating observer

What then? It could be related to the phenomenon suggested in the 1970s by the Canadian physicist William Unruh and others. Combining the ideas of relativity and quantum mechanics, they said that an accelerating observer should see radiation coming at them out of the vacuum.

The astrophysicist Bernard Haisch of the Max Planck Institute for Extraterrestrial Physics in Garching, Germany, and the electrical engineer Alfonso Rueda of California State University in Long Beach realized that the vacuum's interaction with an accelerating body would permeate its entire volume. This could produce a force that acts in the opposite direction to the body's movement. They originally likened it to the way in which charged particles moving through a magnetic field experience a force – the Lorentz force – that affects their motion. In this case, there were electromagnetic interactions with the quantum vacuum.

Mike McCulloch of the University of Plymouth, UK, now thinks such interactions are also just what you need to break the equivalence principle. Unruh radiation should come in a spectrum of many different wavelengths. For very small accelerations, it would be dominated by very long wavelengths. Make the acceleration very small indeed, and some of these wavelengths become longer than the size of the observable universe, effectively cutting them off. In this case, according to calculations McCulloch made in 2007, the total amount of Unruh radiation experienced by a body would drop, and it would feel less of an opposing force. Its inertia would thus fall, making it easier to move than Newton's standard laws of motion dictate – and cutting the connection with gravitational mass.

Anomalous motions

The problem with this idea is testing it. In the high-gravity environment of the Earth, accelerations small enough for the effect to be observed would not be easy to manufacture. But its effects might well be seen in a low-gravity environment such as that found at the edge of a galaxy. Indeed, looking at the anomalous motions of most spiral galaxies, McCulloch suggests that this mechanism could also explain another enduring cosmic mystery – that of dark matter.

It is fair to say that such ideas have not set the world alight. When Haisch and Rueda came up with their mechanism, NASA was sufficiently impressed to fund further study and the duo also attracted some $2 million in private investment. But the lack of testable predictions of how the effect might manifest itself led the money and interest to dry up.

In 2010 three Brazilian astronomers, led by Vitorio De Lorenci of the Federal University of Itajubá, suggested one new test. If you use a spinning disc to cancel out the accelerations produced by the Earth's rotation and its movement through space, at minuscule accelerations the disc's inertia would drop, so it would spin faster than expected from Newton's laws. Despite a relatively modest cost, however, no money has yet been forth-coming to fund the experiment.

The deadlock remains, until someone delivers either an experiment that exposes the equivalence principle as a sham, or a theoretical idea that shows why it must be just so. If gravitational mass is indeed inertial mass in another guise, then gravity is truly an illusion that springs from the warping of space, as described by general relativity. Quantum theories of gravity, including string theories, might find themselves laid upon the sacrificial altar instead.

Is everything made of numbers?

When Einstein completed his general theory of relativity in 1916, he looked down at the equations and discovered an unexpected message: the universe is expanding. Einstein did not believe that the physical universe could shrink or grow – so he ignored what the equations were telling him – but 13 years later Edwin Hubble found clear evidence of the universe's expansion. Einstein had missed the opportunity to make the most dramatic scientific prediction in history.

How did Einstein's equations 'know' that the universe was expanding, when he did not? If mathematics is nothing more than a language we use to describe the world, an invention of the human brain, how can it churn out anything beyond what we put in? 'It is difficult to avoid the impression that a miracle confronts us here', wrote physicist Eugene Wigner in 1960.

The prescience of mathematics seems no less miraculous today. In 2012, at the Large Hadron Collider, physicists observed the fingerprints of a particle that had been glimpsed 48 years earlier lurking in the equations of particle physics. How can mathematics know about Higgs particles or any other feature of physical reality? It may be because mathematics is reality, says physicist Brian Greene of Columbia University, New York. Perhaps if we dug deep enough, we would find that physical objects like tables and chairs were ultimately not made of particles or strings, but of numbers.

What might this mean? A starting point is to ask what mathematics is made of. The late physicist John Wheeler said that the basis of all mathematics is $0 = 0$. All mathematical structures can be derived from something called the empty set, the set that contains no elements. Say this set corresponds to zero; you can then define the number 1 as the set that contains only the empty set, 2 as the set containing the sets corresponding to 0 and 1, and so on. Keep nesting the nothingness like invisible Russian dolls and eventually all of mathematics appears. Mathematician Ian Stewart of the University of Warwick, UK, calls this 'the dreadful secret of mathematics: it's all based on nothing'. Reality may come down to mathematics, but mathematics comes down to nothing at all.

That may be the ultimate clue to existence – after all, a universe made of nothing requires no explanation. Indeed, mathematical structures seem not to require a physical origin at all. A dodecahedron was never created, says Max Tegmark of the Massachusetts Institute of Technology. A dodecahedron does not exist in space or time at all – it exists independently of both. Space and time themselves are contained within larger mathematical structures, he adds. These structures just exist; they cannot be created or destroyed.

That raises a big question: why is the universe made of only some of the available mathematics? It is true that seemingly arcane and apparently unphysical mathematics does sometimes turn out to correspond to the real world. Imaginary numbers, for instance, were once considered totally deserving of their name, but are now used to describe the behaviour of electric circuits and elementary particles; non-Euclidean geometry eventually showed up in Einstein's description of gravity. Even so, these phenomena represent a tiny slice of all the mathematics out there.

Not so fast, says Tegmark. He believes that physical existence and mathematical existence are the same, so any structure that exists mathematically is also real. What about the mathematics our universe doesn't use? Other mathematical structures correspond to other universes, Tegmark says. He calls this the level-4 multiverse, and it is far stranger than the multiverses that cosmologists often discuss. Their common-or-garden multiverses are governed by the same basic mathematical rules as our universe, but universes within Tegmark's level-4 multiverse operate with completely different mathematics.

All of this sounds bizarre, but the hypothesis that physical reality is fundamentally mathematical has passed every test. If reality is not, at bottom, mathematics, what is it? Maybe some day we shall encounter an alien civilization and show them what we have discovered about the universe, Greene says. 'They'll say, "Ah, math. We tried that. It only takes you so far. Here's the real thing".'

Interview: It's amazing what our puny brains can do

Sean Carroll, a theoretical physicist at the California Institute of Technology, discusses why physics is on a winning streak at the moment and speculates on the next big breakthrough.

Are you enjoying the current popularity of physics that's come as a result of discoveries like the Higgs boson and gravitational waves?

It's interesting, because physicists sort of ruled the twentieth century with quantum mechanics, the atomic bomb and all kinds of technologies. We had the most political power and intellectual heft. Now the biologists are stealing that from us. Biology is advancing enormously quickly, and has a much more direct impact on our lives. But such advances – gene editing, for example – can be double-edged swords. In a sense, this works in favour of physics: the kinds of discoveries we're making now don't have immediate implications for technology or our everyday lives. No one's worried about how the Higgs boson or gravitational waves are going to be used – they're just really cool.

These physics breakthroughs have come from proving mathematical theorems. Should we continue to use maths to guide research?

It's not just that mathematics is helpful in understanding nature; it's the scientific methodology, too. The bigger point is that these things illustrate the knowability of our world. There's a quiet debate between people who think nature is fundamentally mysterious versus those who think it is fundamentally intelligible. These kinds of discoveries remind us is that our puny little brains have the power to make amazing predictions about faraway and very difficult-to-access aspects of the natural universe.

So what's next in this 'decade of discovery'?

It's impossible to say. We could find proof of cosmic inflation in the early universe, discover dark matter and find some particle that's outside the standard model of physics. Any of those could happen in the next two years.

A hundred years passed between the theory of gravitational waves and their discovery. Do we need to give today's frontier ideas more time?

Absolutely. There's a small part of the human intellectual portfolio devoted to these big, ambitious questions, and you have to let the people who devote themselves to tackling them take their time to work it out. The discovery of gravitational waves by the LIGO collaboration is incredibly impressive for so many reasons: it's not just the number of people, but also the number of years it took.

People started taking the detection of gravitational waves seriously in the 1980s and they knew before they built the first gravitational wave observatory that it probably wouldn't be sensitive enough to see anything – and indeed it didn't. I would give infinite credit to the visionaries who knew this stuff but would not give up, who devoted their lives to making it happen.

Do you hope that your work will encourage the next Einstein?

I don't like to talk about the next Einstein: the large majority of theoretical work is collaborative these days.

9
Conclusion

As Einstein once wrote, 'The most incomprehensible thing about the universe is that it is comprehensible.' Thanks not least to him, our understanding has progressed in leaps and bounds over the past century or so. A few basic principles underlie Einstein's relativity, but these create other, currently undecipherable problems. Some new solutions will be needed to complete the picture. Here is an outline.

PRINCIPLES

Principle 1
THE SPEED OF LIGHT IS A CONSTANT
Nothing can exceed this cosmic speed limit.

Principle 2
THE EQUIVALENCE PRINCIPLE
Gravity and acceleration always look the same.

Principle 3
THE COSMOLOGICAL PRINCIPLE
The universe is the same in all places and in all directions.

SPECIAL RELATIVITY

$E = mc^2$

GENERAL RELATIVITY

THE STANDARD MODEL OF COSMOLOGY

GRAVITATIONAL WAVES

THE COSMIC MICROWAVE BACKGROUND

PROBLEMS

PROBLEMS

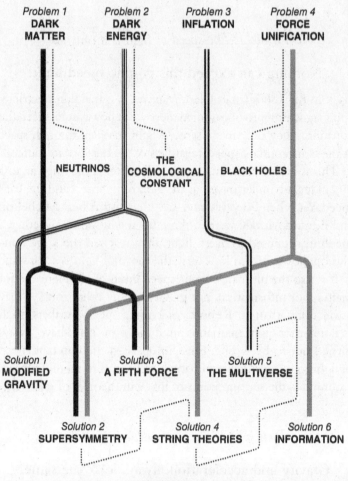

Problem 1
DARK MATTER

Problem 2
DARK ENERGY

Problem 3
INFLATION

Problem 4
FORCE UNIFICATION

NEUTRINOS

THE COSMOLOGICAL CONSTANT

BLACK HOLES

Solution 1
MODIFIED GRAVITY

Solution 3
A FIFTH FORCE

Solution 5
THE MULTIVERSE

Solution 2
SUPERSYMMETRY

Solution 4
STRING THEORIES

Solution 6
INFORMATION

SOLUTIONS

PRINCIPLES

Principle 1: The speed of light is a constant

Nothing can exceed this cosmic speed limit.

Back in the 1860s James Clerk Maxwell was melding electricity and magnetism into one unified theory. But however he sliced the equations, they only made sense if light travelled through space at the same constant speed, regardless of the speed of its source.

This is odd. If someone fires a bullet from a moving car, to a bystander the bullet travels at the sum of its speed and the car's speed. Yet when 20 years later US physicists Albert Michelson and Edward Morley were looking for the luminiferous ether, a medium supposed to carry light, they reached the same conclusion: however you look at it, the speed of light is a constant.

It is also the ultimate cosmic speed limit. No influence – not matter, not information, not gravity or any other force – may travel faster than it. Reports of cosmic speed breakers, such as faster-than-light neutrinos announced in 2011, have always turned out to be wrong. Einstein declared the constant speed of light to be a principle of nature, and began to rebuild physics around it – the starting point of his twin theories of relativity.

Principle 2: The equivalence principle

Gravity and acceleration always look the same.

In the sixteenth century Galileo noticed that falling objects accelerated at the same rate regardless of their mass. A feather and a hammer dropped from the Leaning Tower of Pisa will hit the ground at the same time, once you discount air resistance.

During the *Apollo 15* lunar landing, astronaut David Scott confirmed that principle on the airless Moon.

Newton showed that this could only be true if an odd coincidence held: inertial mass, which quantifies a body's resistance to acceleration, must always equal gravitational mass, which quantifies a body's response to gravity. There is no obvious reason why this should be so, yet no experiment has ever prised these two quantities apart. As with light's constant speed, it was Einstein who declared this equivalence a principle of nature.

Principle 3: The cosmological principle

The universe is the same in all places and in all directions.

A few decades before Galileo, Copernicus dared to suggest that the Earth was not a special place in the cosmos. A century or so later, Newton, in his great treatise *Principia*, assumed that the solar system was embedded in a uniform space that extended vast distances in all directions.

These are the origins of what in modern cosmology has morphed into the cosmological principle: gaze out into the universe and everything is more or less the same everywhere and in whatever direction you look. Local clumps of matter exist in the form of solar systems, galaxies and clusters of galaxies but, on a big enough scale, everything averages out to uniformity.

It's a simplification that makes the mathematics a lot easier when we're trying to build a working model of the cosmos. But our limited view makes it difficult to say whether it truly is a universally valid principle. The discovery of ever-bigger structures, for example, 2013's 10 billion-light-years-wide arc of galaxies dubbed the Hercules–Corona Borealis Great Wall, is calling that into question.

Special relativity

As Einstein worked out, the principle of a constant speed of light has some odd consequences. In everyday experience, two cars approaching each other at 100 kilometres an hour would collide at double that, 200 km/h. But imagine you're sitting in one of two spaceships approaching each other, each travelling at 90 per cent of the speed of light, c. From the perspective of one, at what speed is the other approaching?

The exact figure doesn't matter★ – but it can't be bigger than c. In Einstein's special theory of relativity of 1905, time and space warp to accommodate light's speed limit. Moving clocks tick more slowly and moving rulers appear shorter, so there is no one objective measure of time and space – and you really will age less in a speeding spacecraft. At our normal speeds, these warping effects are negligible, but close to light speed they become hugely significant, and ensure that no object can ever cover a given space in a shorter time than light can. (★It's 99.4 per cent of the speed of light.)

General relativity

Einstein's warped theory of gravity

If motion warps space and time (see 'Special relativity', above), then so does acceleration – and gravity is a form of acceleration. That's the lesson of Einstein's magisterial general theory of relativity of 1916, which combines special relativity and the equivalence principle into our working theory of gravity. Massive objects bend space and time around them, making things seem to accelerate towards them. General relativity provides a framework to explain the large-scale workings of the universe, but a cosmological model requires further information: how matter is distributed.

Gravitational waves

These ripples in space-time were the last unconfirmed prediction of general relativity until they were finally spotted in September 2015. The signal of two massive black holes spiralling together and merging was a triumph of painstaking, patient detective work by the Advanced LIGO experiment.

$$E = mc^2$$

This most famous equation of physics stems from special relativity, and says that mass is just a concentrated form of energy, connected by the constant speed of light. So slam particles together at very high energies, as in CERN's Large Hadron Collider, and you can create other, more massive particles – a path of discovery that eventually led to the standard model of particle physics.

The standard model of cosmology

When Einstein first used general relativity to build a cosmic model, he followed the orthodoxy of the day and assumed that the universe was static: neither expanding nor contracting.

Observations in the 1920s, however, showed that distant galaxies are 'redshifted' as if they are moving away from us. Others then used his theory, plus the simplifying cosmological principle that the universe's matter is uniformly distributed, to build models of an expanding universe. This was the beginning of today's standard cosmological model. It describes a universe that began in the hot, dense, infinitesimal pinprick of the Big Bang some 13.8 billion years ago – and contains a few surprises that we still find hard to explain.

The cosmic microwave background

Discovered by accident in 1964 as a background hiss in a gigantic telecoms receiver, this cool sea of radiation is now seen as clinching evidence of the Big Bang. The oldest light in the universe, it was sent on its way some 380,000 years after the Big Bang, when the cosmos had cooled enough for the first atoms to form, allowing photons to travel freely. Probes collecting this light, most recently ESA's Planck mission, have mapped it in fine detail, providing information on the universe's earliest years and its make-up today.

PROBLEMS

Problem 1: Dark matter

Galaxies rotate too quickly for their visible matter.

The Earth whirls around the Sun at a speed determined by its distance, the Sun's mass and gravity's strength, which is a universal constant of nature. You don't need general relativity to work that out: Newton's old-school gravitation will do. The same law should apply to distant galaxies swirling around their common centre of mass.

In the 1930s, however, the astronomer Fritz Zwicky discovered that the outer parts of the Coma galactic cluster were rotating far faster than the cluster's estimated mass allowed. In the 1970s, Vera Rubin confirmed the huge mismatch in a clutch of spiral galaxies similar to the Milky Way. She estimated that they must contain about six parts of invisible matter for every one part of visible matter.

This 'dark matter' must interact gravitationally to produce the motion, but hardly at all through the other forces of nature. The standard model of particle physics provides no particle that fits the bill, and efforts to detect dark matter particles beyond the standard model, or manufacture them in high-energy particle collisions, have so far come to naught. Something is missing.

Problem 2: Dark energy

The universe is flying apart faster and faster.

In the late 1990s, two groups studying far-off supernovae discovered that these stellar explosions were consistently fainter than expected. Their conclusion was that the space their light had travelled through to get to us had stretched more than expected, so the supernovae were further away than supposed.

Dark energy is the name for whatever is causing this accelerating expansion. It dominates the cosmos, making up some 68 per cent of everything there is. But what is it? Perhaps it is a vacuum energy of the sort that quantum particles might create by popping in and out of free space. This would be a resurrection of the cosmological constant that Einstein introduced into the equations of general relativity, and then dropped. Or perhaps dark energy is a 'quintessence', an as yet undiscovered fifth force of nature.

Both identities have their problems, and there could be another way out. A universe with a variable density of matter would expand at different rates in different places, possibly producing an illusion of accelerated expansion. So if we drop the cosmological principle we might possibly get rid of dark energy.

Problem 3: Inflation

Faster-than-light expansion spawns many other universes.

Range your eye across the cosmos, and a couple of features are hard to explain. It is almost geometrically 'flat', and even far-off bits all have almost the same temperature.

Cosmic inflation solves these problems at a stroke. In its earliest instants, the universe expanded faster than light (light's speed limit applies only to things within the universe). That ironed out wrinkles in its early chaotic self and meant that even now far-flung parts were once in close contact, so could swap heat.

In 2014 researchers claimed to have seen ripples from inflation imprinted on the cosmic microwave background. But this proved mistaken, and it's not clear what would have made the early universe inflate anyway. Worse, inflation is very difficult to stop, creating a multiverse of causally disconnected universes that eternally bud off from one another.

One way out might be to weaken the constant speed of light. If the speed of light were faster in the early universe, this would also explain the temperature problem. Perhaps light is still slowing now, just at a rate that is imperceptible even to our most sensitive detectors.

Problem 4: Force unification

Our theories of reality don't get along.

Gravity is the only force of nature we can't describe by quantum theory. In addition, any effort to make it quantum – to describe it through the exchange of particles called gravitons, rather than through general relativity's space-time warpings – is ripped apart by uncontrollable infinities that render all calculations meaningless. That's a big problem for physicists. When subatomic particles interact, gravity is generally so weak that it can safely be ignored. But in some realms, the two must come together: in black holes, for example, or in describing the universe's tiny origins in the Big Bang. Without a quantum theory of gravity – a first step to a unifying theory of everything – science faces an impenetrable barrier to ultimate enlightenment. Besides, any theory of quantum gravity will require breaking the link between gravitational and inertial mass embodied by the equivalence principle – undermining a foundation stone of modern physics.

Neutrinos

Could neutrinos be dark matter? The standard model of particle physics says that these elusive particles have no mass, but experiments now say they do have a small one – the only direct contradiction of a standard model prediction so far. But the extra mass seems unlikely to be enough to explain dark matter, unless as yet undiscovered new varieties of 'sterile' neutrino exist. Recent results from the European Space Agency's Planck satellite and the Ice-Cube Neutrino Observatory in Antarctica seem to discount that.

The cosmological constant

When Einstein created his static universe model (see 'General relativity' in the Principles section of this chapter), he added an extra term in the equations of general relativity to counteract gravity's pull. He later called this cosmological constant his 'greatest blunder'. Tweaked to represent a quantum energy of free space, it might explain dark energy – but quantum theory supplies a huge 10,120 times as much as is needed to set the universe speeding on its way. This is perhaps the most glaring numerical mismatch in all of physics.

Black holes

Black holes are super-dense objects that swallow everything, including light, that strays too near. They come in different sizes: supermassive black holes lurk at the heart of most galaxies, and stellar-mass ones form when spent stars collapse in on themselves. Predicted by general relativity, black holes are also places where gravity is so strong that it can no longer be neglected on quantum theory's small scales – so we currently have no understanding of what happens at the edge of a black hole or inside one.

SOLUTIONS

Solution 1: Modified gravity

Our theories of gravity have only ever been tested on small scales.

General relativity is a hugely accurate theory of gravity – as far as we can tell. But could modifying gravity perhaps exorcise some cosmic demons?

General relativity's predictions of the movements of probes and planets are accurate up to the scale of the solar system, and the recent discovery of gravitational waves indicates that they are also correct about what happens when two orbiting black holes merge. But by cosmic standards, these systems have a lot of mass crammed into a relatively small space. What happens to gravity in environments where its strength is weaker?

A tweaked version of gravity called modified Newtonian dynamics (MOND) might describe the anomalous galactic rotations taken to be evidence for dark matter, but there is as yet no theory for how this modification might come about. The DGP gravity hypothesis, meanwhile, named after the initials of its originators, suggests that gravity can leak out from our 4D space-time into a higher-dimensional 'bulk', progressively weakening its effect over time and so producing the illusion of dark energy. We're yet to find any experimental evidence to back up this hypothesis, however.

Solution 2: Supersymmetry

More particles can explain why the universe is as it is.

Supersymmetry, or SUSY to its friends, is the Swiss army knife of particle theories: it has a tool for everything. It works by suggesting that for every matter-making fermion in the standard model there is a force-carrying boson, and vice versa.

SUSY was originally formulated to solve problems in particle physics such as why the other three forces of nature besides gravity have such different strengths. But it also turned out that the lightest of the 'superpartner' particles it proposed, the neutralino, supplied a ready-made identity for dark matter.

This is probably too good to be true, though: if supersymmetric particles exist, they should have been created at CERN's Large Hadron Collider by now. But aside from the odd exciting blip that subsequently went away, there's been no sign. Perhaps the superpartners are heavier than we thought, or perhaps we just haven't teased their existence from the data yet. But increasingly, particle physicists are facing an uncomfortable thought: nature might not have supplied as neat an answer as supersymmetry.

Solution 3: A fifth force

Could a quintessence banish cosmic ghosts?

Gravity, electromagnetism, the weak force and the strong force … four seems an arbitrary number of fundamental forces. Why not more?

The most likely fifth force is a weak, long-distance force, a bit like gravity, that would probably interact with it. It might cancel a little of it and explain why the universe's expansion rate is accelerating, mimicking dark energy. Or it could add to gravity to explain the additional pull usually attributed to dark matter. Such a fifth force would have to be cunningly disguised to explain why we don't feel it. One suggestion is that the large amount of mass in and around the solar system shields us from its effects. That also makes it a very hard idea to test.

Fifth forces with attendant quantum particles are also sometimes proposed to solve problems such as fine-tuning in particle physics, but there seems very little evidence for new forces on this scale. The occasional anomaly does keep physicists hoping, however: most recently, aberrations in the expected decay rate of radioactive beryllium nuclei have been touted as evidence of a whole new 'dark sector' of particles and forces.

Solution 4: String theories

An ultimate theory must subsume quantum theory and relativity.

Many physicists' dreams, including Einstein's, have been dashed in attempts to formulate a unifying theory of all of nature's phenomena. In the past few decades, a popular route to a theory of everything has emerged: string theory, and in particular a variant known as M-theory. In M-theory, matter is made not of pointlike particles, as in the standard model, but of one-dimensional vibrating strings living in a universe with 11 space-time dimensions. These strings vibrate in different ways to create various elementary particles – even gravitons – to carry a quantized gravitational force.

M-theory could solve fine-tuning and implicitly includes supersymmetry. Its extra dimensions are curled up and tiny, explaining why we are not aware of them. Breakthroughs such as the discovery in 1995 of a way to link a theory with gravity in five dimensions with a purely quantum theory in four seemed to indicate that it was the right way to unify forces and theories.

But advances similar to this 'AdS/CFT correspondence' are yet to be made for space-times more like that of our universe, and string theory has yet to make a single testable prediction. Most perturbingly, it seems to predict the existence of a multiverse of anything up to 10,500 different universes (see below).

All this has led some to question whether string theory counts as science. Rival approaches do pop up from time to time. But all these theories are far from a theory of everything as most people would understand it: one that can also explain, for example, how properties such as consciousness 'emerge' from the workings of inanimate matter.

Solution 5: The multiverse

The universe is as it is – because every other universe is out there too.

Many roads lead to the multiverse. String theory needs it. Inflation creates it. In the many-worlds interpretation of quantum mechanics, we are constantly, unwittingly, making parallel universes. The first difficulty is that these multiverses are probably all different. The second is how you get convincing evidence for the existence of any of them.

Multiverses are in general both a blessing and a curse. The string theory or inflationary multiverse, for example, can solve problems like fine-tuning – the idea that our universe is fine-tuned for life. Other universes exist where all other possible configurations of matter exist, and our configuration just happens to be one of those where conditions were ripe for sentient, questioning beings to evolve. But such 'anthropic' reasoning absolves us of the responsibility of asking the most probing of questions, 'Why?'. It may be the best we can do, but by allowing every possibility besides the one you're probing to play out somewhere in the multiverse, science robs itself of its predictive power.

Solution 6: Information

Energy and matter don't matter: information is where it's at.

When attempting to unify general relativity and quantum theory, it is generally assumed that general relativity is at fault. It is, after all, a classical field theory of the sort that newer quantum theories in the twentieth century largely nudged aside.

General relativity starts from the assumption that matter, energy and space-time are fundamental building blocks of the universe. But what if that's not the case? A common theme among researchers trying to look beyond general relativity and quantum theory to a more unified understanding of nature is that something else lies at the root of all things: information. Perhaps looking at the universe through information-tinted spectacles will open us up to blindingly obvious solutions that make the problems we encounter today melt away.

Where next for relativity?

'The difference between stupidity and genius is that genius has its limits', said Albert Einstein. Relativity, too, must have its limits. When we look into the heart of a black hole, back to the beginning of time or down to the finest scales of reality, we know that Einstein's theory cannot give us all the answers. One day, some still-more-powerful model of gravity, space and time will supplant it, whether woven from loops or strings or membranes or something as yet unimagined, perhaps bringing as profound a shift in our picture of the universe as Einstein's own revolution. Our understanding of the universe has been completely rewritten in the past century – and is likely to be rewritten again.

While we know those limits are out there, relativity keeps growing in stature and scope. With the discovery of gravitational waves, we have not only found that another of its startling theoretical predictions turns out to be real, but also gained an entirely new way to explore the universe. The warping of time and space, once an outlandish idea, must now be routinely calculated in order for satnav to work, while atomic clocks promise to exploit gravitational time dilation to measure height. And we are only just learning to solve Einstein's fearsome equations using computer simulations, to predict what happens when black holes or neutron stars collide, and to find out whether the space-time dimples caused by galaxies could change the fate of the universe.

More than a century old and still not fully understood, relativity is almost certainly going to spring some surprises on us in the years ahead.

Six spots for space-time tourism

1 **Ulm**, Germany: The house where Einstein was born is next to the railway station, the equations of relativity appear in a stained-glass window of the Lutheran church known as the Minster, and other monuments to the great man exist in this southern German city.

2 **Bern**, Switzerland: This is home to the patent office where Einstein served as Technical Assistant (level III) and the Einsteinhaus, a museum in the house where he and his family lived from 1903 to 1905.

3 While in **Switzerland**, you could seek out an Einstein-era train, the kind of vehicle that inspired his experiments involving light and clocks, leading to the disturbing conclusion that even Swiss trains can't be punctual to all observers. There's the Blonay-Chamby railway museum, for example.

4 **Príncipe**: On this tropical island off the west coast of Africa, Arthur Eddington made the first confirmation of the predictions of general relativity by observing the position of stars during a total eclipse.

5 **Washington DC**: In the grounds of the National Academy of Sciences you can visit the Albert Einstein Memorial statue, and even sit in his lap if you are so inclined.

6 **V616 Monocerotis**: If your budget stretches to an interstellar starship (and uploading your consciousness to its computer), then in a few millennia you could find out what really happens when space-time is stretched to breaking point by visiting the nearest known black hole, which is about 3,000 light years away.

Nine references to relativity in music, movies, literature and art

1 For that authentic 1980s-nerd-party feel, check out synthpop hit **'Einstein A Go Go'**, by British band Landscape.

2 More recently, Kelly Clarkson dazzled lyric lovers with the profound line 'dumb plus dumb equals you', in her song called simply **'Einstein'**.

3 The great mind has inspired some more highbrow music too. *Einstein on the Beach*, an opera by Philip Glass, is named after a famous picture of … Einstein on the beach.

4 In *Relativity Rag*, composer George Benjamin experiments with warping the familiar musical form in a way supposedly influenced by Einstein's theory.

5 The film *Interstellar* features an exceptionally realistic fake black hole, simulated by relativity expert Kip Thorne of the California Institute of Technology.

6 In the film and play *Insignificance*, Einstein becomes entangled with Marilyn Monroe, Joe DiMaggio and Joseph McCarthy as they all meet in a Manhattan hotel room.

7 For a more serious attempt at portraying the man and his ideas, try the BBC drama-documentary *Einstein and Eddington*.

8 Salvador Dalí's painting *The Persistence of Memory* shows the influence of relativity in its warped, melting watches.

9 M. C. Escher also inhaled the heady air of relativity for his student-hall-ubiquitous print that plays with gravity and space, *Relativity* (the one with the stairs; not the other one with the stairs).

Ten deep thoughts by Einstein

1   'The most incomprehensible thing about the world is that it is comprehensible.'

2   'If A is success in life, then A equals x plus y plus z. Work is x; y is play; and z is keeping your mouth shut.'

3   'According to the general theory of relativity, space is endowed with physical quantities; in this sense, therefore, an ether exists. Space without an ether is inconceivable.'

4   'Space has devoured ether and time; it seems to be on the point of swallowing up also the field and the corpuscles, so that it alone remains as the vehicle of reality.'

5   'What I'm really interested in is whether God could have made the world in a different way; that is, whether the necessity of logical simplicity leaves any freedom at all.'

6   '… my intellectual development was retarded, as a result of which I began to wonder about space and time only when I had already grown up.'

7   'Creating a new theory is not like destroying an old barn and erecting a skyscraper in its place. It is rather like climbing a mountain, gaining new and wider views, discovering unexpected connections between our starting point and its rich environment. But the point from which we started out still exists and can be seen, although it appears smaller and forms a tiny part of our broad view gained by the mastery of the obstacles on our adventurous way up.'

8   'All our science, measured against reality, is primitive and childlike – and yet it is the most precious thing we have.'

9 '... the distinction between past, present and future is an illusion, although a persistent one.'

10 'A man should look for what is, and not what he thinks should be.'

Eight anecdotes, jokes, facts and myths

(to be used in party conversation with appropriate caution)

1 It is a common myth that Einstein was a poor student. This idea may have emerged because the grading system at the time went up to 6, not 10. His matriculation certificate (http://rarehistoricalphotos.com/albert-einsteins-matriculation-certificate-1896/) shows that at the age of 17 he was getting good-to-excellent grades in everything … except French.

2   *There was a young lady named Bright,*
    Whose speed was far faster than light.
    She set out one day
    In a relative way,
    And returned home the previous night.

3 When early-adopter relativity expert Ludwig Silberstein approached Arthur Eddington at a party in 1919, he suggested that Eddington might be one of the three men who actually understood the general theory of relativity. Eddington was slow to reply. 'I was wondering who the third one might be', he eventually confessed, to a presumably downcast Silberstein.

4 Einstein loved a power nap, apparently.

5 A joke:

The barman says, 'Sorry, we don't serve hypothetical faster-than-light particles in here.'
A tachyon walks into a bar.

6 David Ben-Gurion offered Einstein the chance to be Israel's first President. He declined.

7 Despite perceptions that he was a kindly old gent, Einstein was capable of being quite belligerent, and even unreasonable. In July 1936 he replied to editors at the *Physical Review* (who had in the usual fashion sent his submitted paper out for peer review) in these words: 'Dear Sir, We (Mr Rosen and I) had sent you our manuscript for publication and had not authorized you to show it to specialists before it is printed. I see no reason to address the – in any case erroneous – comments of your anonymous expert. On the basis of this incident I prefer to publish the paper elsewhere.'

8 John Wheeler is often said to have coined the term 'black hole', but in fact he adopted it from an anonymous audience member at a talk in 1967: 'In my talk, I argued that we should consider the possibility that the centre of a pulsar is a gravitationally completely collapsed object. I remarked that one couldn't keep saying "gravitationally completely collapsed object" over and over. One needed a shorter descriptive phrase. "How about black hole?" asked someone in the audience. I had been searching for the right term for months, mulling it over in bed, in the bathtub, in my car, whenever I had quiet moments. Suddenly this name seemed exactly right.'

Eight people Einstein corresponded with

1 **Sigmund Freud.** In 1932 Einstein and the psychoanalyst corresponded about violence and war. Einstein wrote, 'Is it possible to control man's mental evolution so as to make him proof against the psychosis of hate and destructiveness?' Freud was sceptical: 'There is no likelihood of our being able to suppress humanity's aggressive tendencies. In some happy corners of the earth, they say … there are races whose lives go gently by: unknowing of aggression or constraint. This I can hardly credit; I would like further details about these happy folk.'

2 **Tyfanny**, a young South African girl. She and Einstein exchanged several letters. In the final one, she wrote, 'I forgot to tell you, in my last letter, that I was a girl. I have always regretted this a great deal, but by now I have become more or less resigned to the fact,' to which Einstein replied, 'I do not mind that you are a girl, but the main thing is that you yourself do not mind. There is no reason for it.'

3 **William Du Bois**, the historian, civil rights activist and founder of National Association for the Advancement of Colored People.

4 **Rabindranath Tagore**, the Indian poet and polymath. They met in Einstein's Berlin home in 1930 and discussed topics such as science and truth and the nature of reality.

5 **Erwin Schrödinger.** Shortly after Schrödinger published his famous paper highlighting the absurdities of quantum mechanics with his simultaneously dead-and-alive cat, Einstein wrote to him, saying: 'From the point

of view of principles, I absolutely do not believe in a statistical basis for physics in the sense of quantum mechanics, despite the singular success of the formalism of which I am well aware.'

6   **President Franklin D. Roosevelt.** In 1939 Einstein co-signed a letter to the president written by the Hungarian physicist Leo Szilard, warning that Germany might develop a nuclear bomb.

7   **Seiei Shinohara**, a philosopher and translator. Shinohara originally wrote to Einstein in 1953 criticizing his role in the development of nuclear weapons, but the two later developed a friendly correspondence.

8   **Eduard 'Tete' Einstein.** Einstein's second son was diagnosed with schizophrenia aged 20 and spent much of his life in psychiatric institutions. Although Albert said to friends that it would have been better if Eduard had never been born, a newly discovered letter from father to son sheds a warmer light on their relationship. Albert writes: 'It seems to me it has been so long since I have seen you, and I am longing to have you around me once again.'

Nine ways to find out more

1 **Einstein online** (www.einstein-online.info), a web portal from Germany's Max Planck Institute of Gravitational Physics (otherwise known as the Albert Einstein Institute), provides a wealth of information about the great man's theories and their applications.

2 *Einstein's Masterwork: 1915 and the General Theory of Relativity* is John Gribbin's 2015 celebration of the theory's beauty and power.

3 *Black Holes & Time Warps: Einstein's Outrageous Legacy* (1994) is physicist Kip Thorne's homage to the great man's ideas, with a foreword by Stephen Hawking.

4 *New Scientist* magazine and its archive at www.newscientist.com contains many Einstein and relativity-related articles going back to 1989.

5 *Dear Professor Einstein: Albert Einstein's Letters to and from Children* (2002) contains many letters from young people and Einstein's replies to them.

6 **einsteinpapers.press.princeton.edu** holds the collected papers of Albert Einstein – a massive written legacy comprising more than 30,000 documents.

7 **www.alberteinstein.info** is another comprehensive digital archive of Einstein's scientific and non-scientific manuscripts, held at the Hebrew University of Jerusalem.

8   *Relativity: The Special & the General Theory*, first published in English in 1920, is a semi-accessible account – with equations but no tensor calculus – by the man himself, which includes a memorable metaphor for spacetime in the form of the flexible reference-mollusc.

9   *Gravity from the Ground Up* (2003) is an introductory guide to gravity and general relativity by Bernard Schutz.

Glossary

Antimatter Every particle has an antiparticle with the same mass but the opposite electric charge. The proton has the negatively charged antiproton; the electron has the positively charged anti-electron, or positron. (There are also antineutrons, even though neutrons have zero charge, as the constituent quarks are charged – so antineutrons are made from antiquarks.)

Big Bang According to the Big Bang Theory – our best explanation for why space is expanding – everything we see exploded from a superhot microscopic region about 13.8 billion years ago.

Black hole A high-gravity object that swallows everything, including light, that strays too near it. Black holes range in size from supermassive holes lurking at the heart of most galaxies to stellar-mass holes that can form when the spent cores of large stars collapse in on themselves.

Boson A subatomic particle with integer spin that carries nature's forces.

Classical physics Physics that predates quantum mechanics and relativity, such as Isaac Newton's laws of motion.

Cosmic Background Explorer (COBE) A NASA satellite that investigated cosmic microwave background (CMB) radiation.

Cosmic microwave background (CMB) A cool sea of radiation discovered in 1964, and the oldest light in the universe. It was sent on its way about 380,000 years after the Big Bang, when the cosmos had cooled enough for the first atoms to form, allowing photons to travel freely.

Cosmological constant An energy density inherent to space, which within Einstein's general theory of relativity creates a repulsive force.

Cosmological principle This states that the universe is more or less the same no matter where you are or in which direction you look.

Dark energy A theoretical form of energy that dominates the cosmos, making up roughly 68 per cent of everything there is, causing the universe to grow at an ever-increasing pace.

Dark matter A mysterious form of matter that comprises about 27 per cent of everything there is, far outweighing ordinary matter and acting as a gravitational glue to form stars and galaxies.

Entropy The degree of disorder in a system.

Event horizon The boundary around a black hole, from which nothing can escape.

Fermion A subatomic particle with half-integer spin, such as an electron or a proton.

General relativity Einstein's 1915 theory combines the ideas of special relativity and the principle of equivalence into a theory of gravity. Objects bend space-time, making things accelerate towards them.

Gravitational wave A ripple in space-time predicted by general relativity, and finally confirmed in 2015.

Graviton A hypothetical particle that would transmit the force of gravity in quantum theory.

Gravity Although dominant over large cosmic scales and near great masses like planets and stars, it is the weakest of the four known forces of nature, and the only one not explained by quantum theory.

Inflation The idea that just after the Big Bang there was a blisteringly fast expansion of the universe. The theory explains several features of the universe, including why it is so flat and smooth.

Hawking radiation The radiation predicted to be released by black holes, due to quantum effects near the event horizon.

LIGO The Laser Interferometer Gravitational-wave Observatory, with twin detectors in Hanford, Washington, and Livingston, Louisiana, which detected gravitational waves for the first time in 2016.

Loop quantum gravity One attempt to make a quantum theory of gravity, in which space-time is formed from tiny loops.

M-theory The modern form of string theory, in which matter is made not of point-like particles but of vibrating strings and membranes living in a universe with 11 dimensions of space-time. These strings and membranes vibrate in different ways to form various elementary particles.

Multiverse A hypothetical multitude of universes. There are different types of multiverse: the inflationary multiverse, for example, arises due to the exponential expansion of space-time, which means that far beyond the edge of the observable universe are countless other bubble universes, inaccessible to us. One interpretation of quantum mechanics suggests the existence of countless universes parallel to our own and interacting to generate quantum phenomena.

Neutron star A remnant of an exploded star that is so dense that atomic nuclei are crushed together and dissolve, leaving a soup dominated by neutrons or perhaps even free quarks.

Principle of equivalence The idea that the effects of gravitational fields are indistinguishable from the effects of accelerated motion.

Photon A particle of light or other electromagnetic radiation.

Quantum mechanics The laws explaining behaviour at the atomic and subatomic level, where particles move like waves, may be in several states at once, and can have shared states connecting them across time and space.

Quarks Building blocks of matter that combine to form composite particles called hadrons, the most stable of which are protons and neutrons.

Redshift The shift in wavelength towards the red end of the spectrum. It can be caused when an object is receding (the Doppler shift) or on a cosmic scale by the expansion of space-time, which stretches light passing through it. The more expanding space the light has passed through, the greater the degree of redshift, so far-off objects appear redder.

Space-time According to relativity, space and time are no longer the absolute and unchanging scaffolding of the universe. Different observers disagree on measurements made in space or time alone; but they do agree on what is happening in the unified space-time.

Special relativity Motion, position and time are all relative, according to Einstein's 1905 theory – all because of the constant speed of light.

Standard model of particle physics This covers the workings of three of the four forces of nature. It describes the interactions of force-carrying boson particles with matter-making fermions according to the mathematics of quantum field theory.

String theory The theory that all particles are manifestations of more fundamental vibrating strings.

Supersymmetry (SUSY) An extension to the standard model of how particles and forces interact, suggesting that, for every matter-making fermion in the standard model, there is a force-carrying boson, and vice versa.

Theory of everything The all-embracing yet elusive theory of physics that unifies quantum mechanics and general relativity, and can describe all the forces of nature in a single framework.

Wormhole A short cut from one part of space-time to another.

Picture credits

All images © *New Scientist* except for the following:
Figure 1.1: Universal History Archive/Universal Images Group/
REX/Shutterstock
Figure 1.2: F&A Archive/REX/Shutterstock
Figure 2.5: ESA/Hubble & NASA
Figure 3.1: NASA/JPL-Caltech
Figure 4.2: LIGO Matt Heintze/Caltech/MIT/LIGO Lab
Figure 5.1: ESA and the Planck Collaboration
Figure 5.4: Daniel Luong-Van, NSF
Figure 6.3: Vogel Springel/Max Planck Institute for
Astrophysics/SPL
Figure 6.6: Steven Jason Saffi
Figure 8.2: Mehai Kulyk/SPL

Index

Note: Page numbers in *italics* refer to photographs or figures.

Interested in learning more?

Learn more about the world and the big issues affecting us by downloading New Scientist Instant Expert audiobooks and ebooks today.

All of the New Scientist Instant Expert audiobooks and ebooks in this series are available to purchase from the Instant Expert app and from instantexpert.johnmurraylearning.com

Use **NSIE40** at instantexpert.johnmurraylearning.com for 40% off any purchase.